CW00688022

The Other Way Around

Frank P Verdon

The Radcliffe Press
London • New York

Published in 1996 by
The Radcliffe Press
45 Bloomsbury Square
London WC1A 2HY

175 Fifth Avenue
New York NY 10010

In the United States of America
and in Canada distributed by
St Martin's Press
175 Fifth Avenue
New York
NY10010

A full CIP record for this book is available from the British Library

A full CIP record for this book is available from the Library of Congress

ISBN 1 86064 030 3

Library of Congress catalog card number: available

Copy-edited and laser-set by The Huguenot Agency, Swindon, England
Printed and bound in Great Britain by WBC Ltd, Bridgend, Mid Glamorgan

Contents

Black and White Plates

Colour Plates

Dedication

For many of the cruises recorded here, the Master of RRS *Charles Darwin* was Captain Sam Mayl who, very sadly for all who knew him, died in April 1989 having been taken ill whilst Master on Cruise CD37/89 in Antarctica. Sam was an active supporter of the Marine Society, and had been working on the idea of an educational project for the Society based upon the global circumnavigation. With that in mind, he had assembled many of the facts and figures which support the text herein; in that respect, he can be said to have been the inspiration for this book.

It is, therefore, with affection and fond remembrance that this book is dedicated to the memory of Sam Mayl.

Sam Mayl 1932 - 1989

Born in Cairo, the son of an army officer, Sam Mayl showed every sign of wishing to make the sea his life. After attending Bristol Cathedral School, he joined Eagle Oil & Shipping Company as a deck apprentice, where he stayed until that company was taken over by Shell. Making steady progress through the ranks, he joined firstly Fishers and then Safmarine, where he attained the rank of Master. He subsequently became Marine Superintendent with the company, but his love affair with the sea encouraged him to join NERC for an initial brief period between 1972 and 1973 before moving to Whitco as a Master again. He rejoined NERC in 1975, serving the remainder of his life as Master of each of the research ships based at Barry, in South Wales.

Sam's primary interests were music and photography, and he was a frequent contributor to *The Marine Observer* with pictures of birds or unusual weather phenomena. Those of us who were privileged to be invited to Sam's cabin were also initially surprised to find that this quiet, well-respected man used to relax at sea by reading French literature in its native language.

Acknowledgements

This book would not have been possible without the active and considerate help of all the scientists and technicians mentioned throughout the text. They were without exception courteous in their criticism of my (at times faltering, at times erroneous) attempts to convey the excitement of their efforts and discoveries in language less esoteric than that of their formal papers, and for that I am very grateful. Nevertheless, the responsibility for the finished product rests with me.

I must also thank those publishers and authors who have allowed me to quote, at times extensively, from their books. I am particularly grateful to Professor Richard Darwin Keynes and Cambridge University Press for their permission to reproduce the map of the voyage of HMS *Beagle*, and the various quotations from the diary of Mr Charles Darwin during that voyage.

Finally, I wish to thank the Natural Environment Research Council, whose support made this book possible.

Frank P Verdon
Barry
1995

Foreword

by Sir Anthony Laughton FRS
Director of NERC's Institute of Oceanographic Sciences
Deacon Laboratory 1978-1988

In 1831 the young and inexperienced Charles Darwin set off in HMS *Beagle* on a voyage around the world that would revolutionise biological thinking and lay the foundations of much of marine geology. Just over one hundred and fifty years later, the Royal Research Ship *Charles Darwin*, one of the research ships owned and operated by the UK's Natural Environment Research Council (NERC), undertook a global circumnavigation in the opposite direction spending three years addressing some of the fundamental problems of today's oceanography and marine geology.

This book is about doing global oceanography, about the planning, the logistical support needed, the politics of international cooperative research, the personal problems and triumphs, and the agonies of catastrophic failure as well as the joys of scientific success.

A global circumnavigation is not lightly undertaken. A hundred years ago HMS *Challenger* made its historic voyage which is taken to mark the beginnings of true oceanographic research. It was not until 1950 that another HMS *Challenger*, of the Hydrographic Department, made a surveying voyage of 75,000 miles around the world carrying out geophysical research at the same time; RRS *Discovery II* made Antarctic circumpolar expeditions in 1932 and 1951, extending its research into the Indian and Southern Oceans in 1950-51; and RRS *Discovery III*, which is still in commission following a major conversion in 1992, ventured into the Indian Ocean in 1963.

Foreword

Unlike the almost informal arrangements that led to the voyage of Mr Charles Darwin in HMS *Beagle*, subsequent circumnavigations required much more detailed planning. Before the *Challenger* expedition of 1872, Governmental support and funds had to be assured

> ... 'a number of factors contributed to the success of the application (to the Government to send out a scientific expedition in 1871), including the support of Richards (the Hydrographer) at a time when the Hydrographic Department was being forced into the deep ocean by the needs of submarine telegraphy, the fear that Britain's lead in deep sea science would be eclipsed by other countries, and particularly by the United States, and even the fact that the Government's financial structure had recently been reformed under Gladstone's Chancellorship. But it is clear that Carpenter's (Vice-President of the Royal Society) personal acquaintance with senior ministers, including Gladstone himself, was of crucial importance'.[1]

With a few name changes these words might equally apply to the planning for RRS *Charles Darwin*, which completed its global circumnavigation between June 1986 and September 1989.

As I emphasised in a keynote speech to mark the return of RRS *Charles Darwin* to the United Kingdom in 1989, the oceans are linked together as part of one large system. They know no natural boundaries. Even though man may define the territorial limits and the boundaries of national jurisdiction, the movement of water and its interaction with the atmosphere, the migration of fish, the dispersion of pollutants, and the movement of tectonic plates are ignorant of these. But the oceans beyond national jurisdiction are used and exploited by man for transport, for defence, for resources, and for recreation, and have an overwhelming effect on all nations.

Oceanography is the scientific study of the oceans in all these aspects:

- the water movements and transport of heat, nutrients and life;
- the interaction of the ocean surface with the atmosphere;
- the biological food chains from the smallest picoplankton to the largest whales;
- the chemistry of seawater;

- the transport and deposition of sediments from the land or from the surface biomass;
- the origin and geological evolution of the ocean basins themselves.

The science of oceanography developed from observations which led to a series of questions. The answers to these questions led to theories and models, (now using highly sophisticated computers) resulting in predictions. These in turn require further observations to test their accuracy and improve the modelling.

The United Kingdom has a long tradition of global oceanography, dating back perhaps not as far as Charles Darwin (who was primarily a geologist and naturalist) but certainly to the voyage of HMS *Challenger* in the 1870s. It was the pressure to continue this tradition in the 1980s, coupled with an increasing recognition that there were many scientific areas that could only be addressed by a cohesive global voyage that led NERC to support the circumnavigation of RRS *Charles Darwin*.

I am delighted that NERC has now decided to record for future generations of marine scientists and all who are interested in the sea the achievements of that circumnavigation, because such events occur only once or twice a century. It is all too easy to forget or overlook the planning and logistics necessary in such an endeavour, and the human efforts and stories that support the success of each and every cruise. Frank Verdon was one of the officers in NERC responsible for overseeing the cruise operation of the expeditions, and is therefore well placed to record in this book the difficulties that had to be overcome, the sense of achievement of both scientists and crew members and the enthusiasm of those who were visiting parts of the world they had never visited before. It is also, in my mind, essential that the international aspect of marine science today, both in the collection of data and in its interpretation and application of that data, be recorded for future generations of scientists and lay persons alike.

I commend this volume to all who are interested in the universality of oceanography.

Chiddingfold, Surrey
1995

---

[1] H L Burstyn *'Science and Government in 19th Century: the Challenger Expedition.' Bulletin de l'Institut Hydrographique* Special Number 2 (1968): pp603-613

Plate 2: Charles Darwin

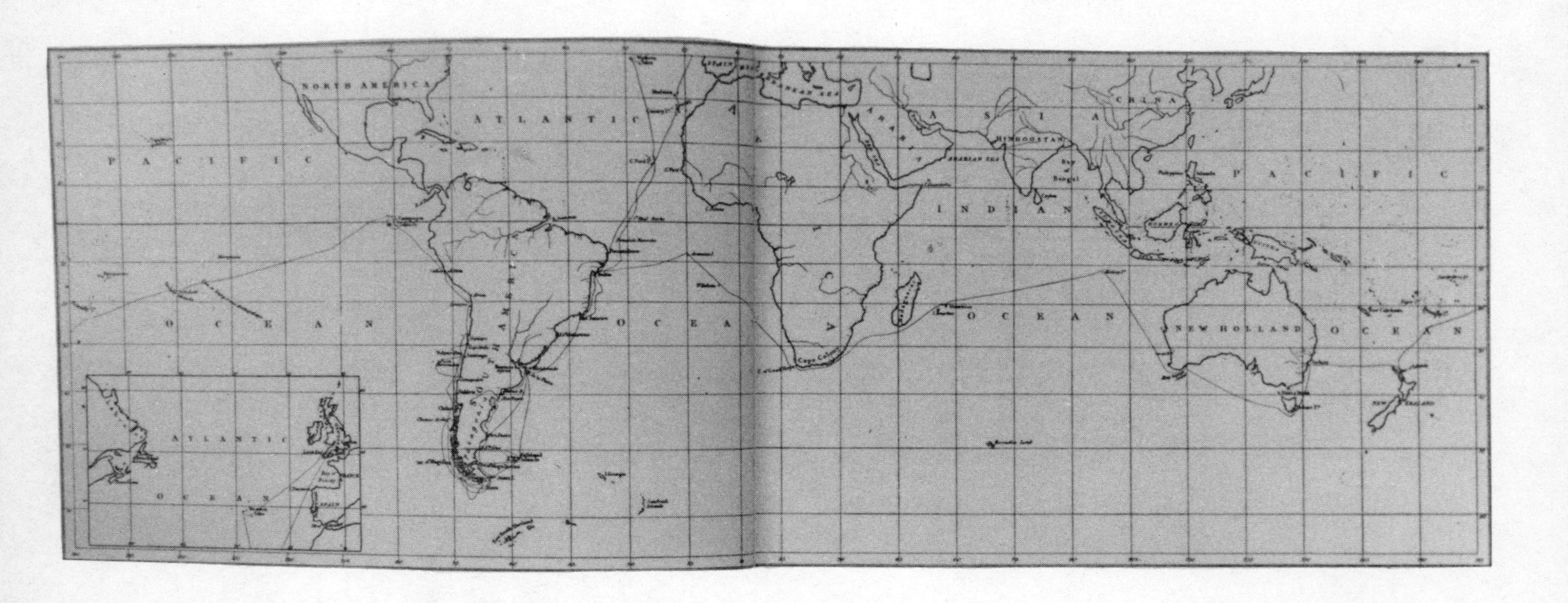

Plate 3: General Chart showing the principal tracks of HMS *Beagle - 1831-1836*

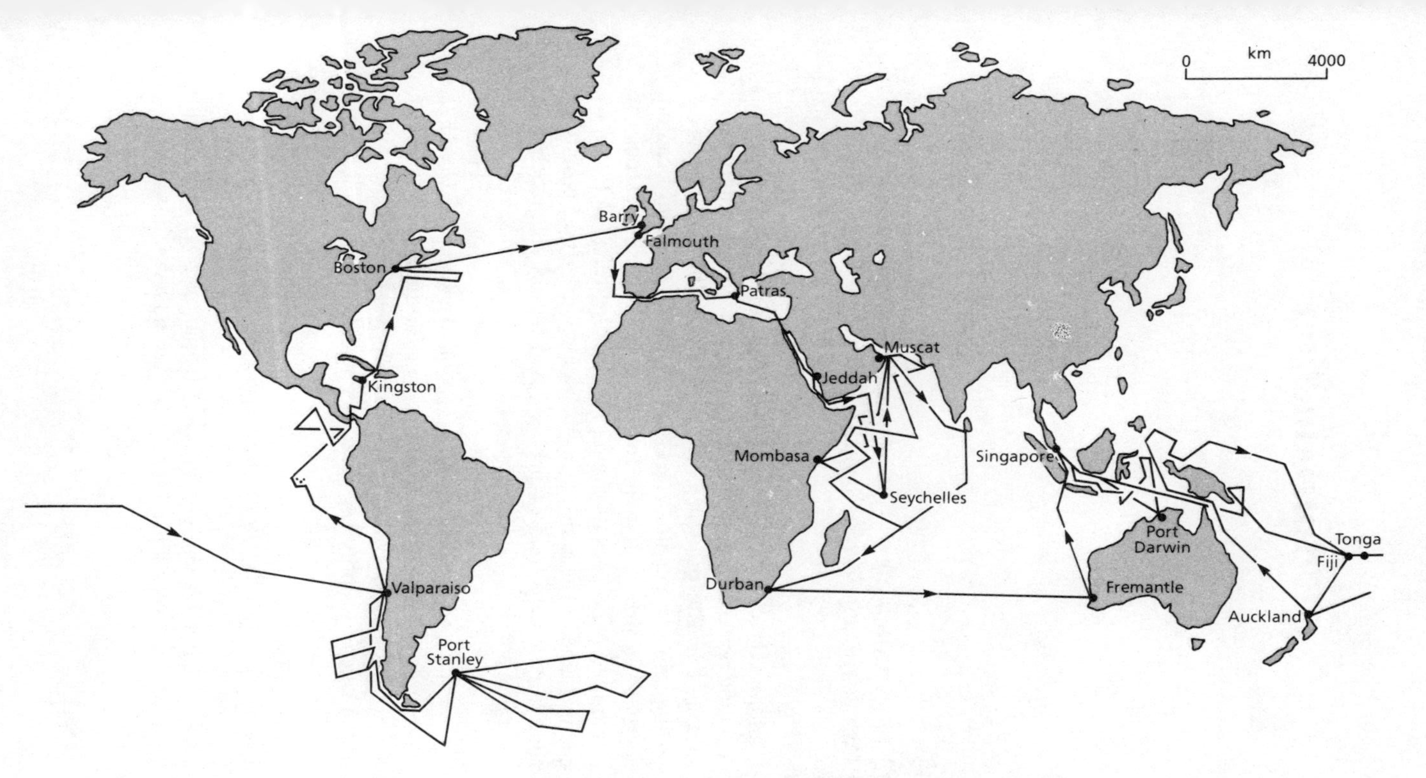

Plate 4: Map of Voyage of RRS *Charles Darwin*

Introduction

In 1831, HMS *Beagle* was recommissioned for survey work, and the Captain (Robert Fitzroy) therefore

> ....proposed to the Hydrographer that some well-educated scientific person should be sought for who would willingly share such as I had to offer in order to profit by the opportunity of visiting distant countries yet little known. Captain Beaufort approved of the suggestion, and wrote to Professor Peacock, of Cambridge, who consulted with his friend, Professor Henslow, and he named Mr Charles Darwin, as a young man of promising ability, extremely fond of geology, and indeed all branches of natural history[2].

At this time, Mr Darwin was 22 years old. He had gone from school to Edinburgh University to study medicine at the age of 16, but had withdrawn from that University after two years. He then went to Christ's College, Cambridge, where he obtained his BA at the age of 21 in 1830.

Professor Peacock wrote to Charles Darwin inviting him to join HMS *Beagle*, about which he (Peacock) wrote

> ...the expedition is entirely for scientific purposes & the ship will generally wait your leisure for researches in natural history etc.

The invite was subject to Darwin's approval by the ship's captain, Fitzroy, and this was nearly the end of the venture. Darwin records in his autobiography

> The voyage of the Beagle has been by far the most important event in my life and has determined my whole career; yet it depended on so small a circumstance as my uncle offering to drive me 30 miles to Shrewsbury, which few uncles would have done, and on such a trifle as the shape of my nose.

(Fitzroy, at this time, was convinced one could tell a man's character by the outline of his features!) Fortunately for science this little problem was overcome, and on 27 December 1831 HMS *Beagle* set sail on her five year voyage with Charles Darwin aboard (at least for much of the time).

The principals involved in the conception of this voyage would all achieve, or had already achieved, fame in the environmental sense. Captain Beaufort, the Hydrographer of the Navy, had devised the 'Beaufort Scale' for recording wind and wave conditions in a shorthand manner. Captain Fitzroy, following the voyage of HMS *Beagle*, was appointed the first Director General of the Meteorological Office, and initiated the system of 'cone' signals for gale warnings around the UK coast. And Charles Darwin's fame needs no further exposition here.

Modern marine science does not depend (at least not overtly!) on personalities to the same degree, and the global voyage of the ship bearing the famous naturalist's name was subject to considerable planning and logistical preparation before it took place. The vessel which was subsequently to be named the Royal Research Ship (RRS) *Charles Darwin* by His Royal Highness the Prince of Wales was ordered by the Natural Environment Research Council in 1982, the centenary of the death of the famous naturalist whose name it carries. Even before the ship was handed over to her owners at the end of 1984, preparations had begun for the global voyage which, unlike that of the man himself, would involve many teams of scientists, a wide variety of disciplines, and significantly more sophisticated scientific equipment than a geological hammer, some 18 notebooks, and a box for sending samples back to England.

Such then is the precursor to the global circumnavigation of the ship, rather than the man. Much has changed in the 150 or so years since Mr Darwin made his voyage, and later his name. It is no longer possible to philosophise about 'geologising' a country, in terms of the difference between applying a geological hammer to granite or soft rocks. Nor does today's scientist spend five years or more away from home, maintaining contact by letter - but then neither does he have the absence of pressure or budgetary freedom enjoyed by this famous naturalist. The most significant contrast between the circumnavigations of the man and the ship may be summarised by just two words - Mr Darwin's efforts were concerned with *qualitative* science, those supported by RRS *Charles Darwin* were totally *quantitative*.

So when did it all begin? That is very much a 'chicken and egg' question, because it had been recognised for some time that the increasingly sizeable and sophisticated oceanographic research equipment required to address the scientific problems of the late 1970s could not be satisfactorily deployed from NERC's existing deep sea vessels. Thus a new vessel, designed to service this need, was required and in April 1982 NERC, through its Director of Scientific Services, Mr Brian F Rule, placed an order for the construction of a new ship, to be named the Royal Research Ship (RRS) *Charles Darwin*. Among the major characteristics of this 70 m (230 ft) vessel were a large clear after-deck, with a 20 tonne 'A'-frame for the deployment and recovery of large equipment, an integrated laboratory suite, single cabins (a novelty at the time), diesel-electric propulsion and a high degree of noise insulation for quiet operation. At the time of her commissioning and for some time thereafter, RRS *Charles Darwin* (see Colour Plate 1) was one of the most advanced research vessels in the world[3].

In fact the primary reason for ordering RRS *Charles Darwin* was not to carry out a global circumnavigation; rather it was to replace the ageing RRS *Shackleton*, which had seen many years of service with both the British Antarctic Survey and Research Vessel Services (RVS). However, the imminence of the new vessel and her facilities sparked enthusiasm in the marine scientific community for something more adventurous than steaming out into the Atlantic or the North Sea, and as early as 1983 the Research Vessel Strategy Committee (RVSC) '...had suggested to the community that proposals would be welcomed for an Indian Ocean Campaign'. Council supported this aspiration at its meeting in January 1984. The perceived groundswell for this initiative derived from the earth scientists, who saw the need which had emerged over the previous decade or two to gather quantitative data in support of the theory of tectonic plates.

It is believed that not all proposals for research started life as altruistic science - one individual who later became a Principal Scientist is reported to have begun his planning by asking the question 'Where would I like to have a holiday?'! Nevertheless, at a meeting called by RVSC, 41 putative principal scientists put forward some 62 proposals for research in the Indian and Pacific Oceans. Council gave approval in March 1985 for the 1986-87 Indian Ocean Campaign. Thus 2-3 years elapsed between the initial concept of an Indian Ocean campaign and its realisation.

In July 1986, shortly after RRS *Charles Darwin* had left Falmouth, NERC's Director of Marine Science (Dr J D Woods) put forward a proposal to continue the voyage eastwards into the Pacific Ocean, and Council approved this. At a meeting of interested scientists in October 1986, specific proposals for a Pacific Ocean campaign were put forward. Not all could form part of a coherent circumnavigation, but the proposals were peer-reviewed and developed into a logistically sound programme which eventually would lead to the completion of the global voyage which had begun earlier in 1986.

Already the comparisons and contrasts are emerging - Charles Darwin received towards the end of August his letter of invitation to join HMS *Beagle*, which was due to sail within a month; the NERC-supported scientists had to put forward proposals which were in gestation for 2-3 years before the campaign began. HMS *Beagle* was at the disposal of Charles Darwin; RRS *Charles Darwin* was to execute a logistically economical voyage consistent with the scientific requirements, but no cruise could be automatically extended because of problems or exciting scientific results. And as already noted, the 'research' of Charles Darwin was essentially qualitative, whereas the remit of the scientists on RRS *Charles Darwin* was to gather data - ie the research was quantitative.

Plates 3 and 4 show the cruise tracks of HMS *Beagle* and RRS *Charles Darwin*, showing that the latter went the other way around.

This book, however, is not solely about comparisons, be they good or bad. It is about the science and scientists supported over a period of three years by NERC and by the crews and technical staff on board the research ship which carries proudly the name of one of the pre-eminent scientists of all time - **Charles Darwin**. When and if comparisons are drawn, as inevitable, between the voyages of the man and the ship bearing the same name, the reader is invited to remember that 150 years have passed between the two circumnavigations, and the naturalist would be as astonished by current research techniques and equipment as we today are astounded by the achievements of a young solo scientist with the limited resources available to him.

Next, a note of explanation is required about the apparently random numbering that occurs throughout this narrative and about the disposition of references throughout the text. For Charles Darwin, the circumnavigation was made on one ship with a sole Captain and a single crew, so any reference to events could be made simply on the basis of date and geography. For the latter-day scientists, their occupation of the ship named after the naturalist was primarily for one or more periods of about 30 days - one cruise. To identify such a period without the necessity to spell out both geographical locations and periods of time, each cruise was given a specific number - eg CD 23/87. The first number is the serial number of the cruise for the ship, starting at 01 for the first cruise following formal commissioning, whilst the second is the year in which that cruise took place. For the global circumnavigation of RRS *Charles Darwin*, the cruise numbers range from CD14/86 to CD41/89.

Two reference systems have been used throughout. The first is that all scientific papers resulting from this global voyage are listed in a References section at the end of the book. When other publications are cited or possibly esoteric terms used, these are explained in chapter end notes.

Finally, to avoid the use of the clumsy he/she, him/her, when I refer to an un-named scientist, I will use the male pronoun throughout. This does not mean that I am unaware or inappreciative of the work done by the ladies - I enjoy their presence as much as any man.

---

[2] R D Keynes, *The Beagle Record* (Cambridge: Cambridge University Press, 1979) xiv+409

[3] For those with an interest in the technical details of the ship, they are found in Appendix 1

1
The Logistics

The multi-role research ships operated by NERC in support of marine science were operated by Research Vessel Services (RVS) based at Barry in South Wales from 1969 until its move to Southampton in 1995. RVS was responsible for the logistical support of the total marine scientific research programme authorised by NERC's Council each year, ensuring that not only did each Principal Scientist go to sea with the equipment essential for his or her research, but that that equipment was properly functioning and supported by technical staff, that all the crew and scientists would be properly fed and watered, that the ships were serviceable with sufficient fuel, lubrication, etc, and, when necessary and perhaps most importantly, that proper formal authority existed for the ship and scientist to carry out research in the territorial waters of other states.

The projected global voyage by RRS *Charles Darwin* was obviously not routine for RVS, but the underlying principles were the same - the ship would have to make portcalls about every 30 days or so for supplies, scientific teams would be joining and leaving at these portcalls, equipment would be transferred on and off the ship, and every 12 months or so there would be a need for a service, what the mariners call a refit. For the normal operational spheres of the NERC ships - primarily the north Atlantic, the Mediterranean, and the North Sea - RVS had built up considerable experience of the port facilities available, and this knowledge was used to complement and supplement the basic information in the handbooks available to mariners. The ports of the

Indian and Pacific Oceans were known only through those handbooks and some limited (and dated) knowledge retained by senior marine staff at RVS from their days before the mast with commercial shipping organisations.

So not without some misgivings, the (then) Marine Superintendent, Capt Michael Perry, persuaded his superiors to allow him to send the RVS Operations Officer, Chris Adams, on a global voyage of his own to most of the ports which preliminary planning had suggested would be used by RRS *Charles Darwin*. The primary purpose of this trip was to observe and record the physical and other attributes of the ports - were there sufficient cranes in the right places to move equipment on and off the ship?; did the port cater for ships as small as RRS *Charles Darwin*?; could it provide the fuel and fresh food required at the right prices?; were there adequate air services and hotels for the personnel movements required, and would these be affected by, say, holiday peaks?; were there any problems not revealed by the handbooks? Other related issues also came into the reckoning, as we shall see later.

At the heart of any call in a port away from home is the ship's agent. He accepts the responsibility for ensuring that things happen as and when they are supposed to. He is the one who will chase Customs if there is a hold-up of equipment at the airport, who will provide transport for weary scientists from their incoming flight to the hotel, at which he will have booked and checked the appropriate number of rooms, who will nominate a provider for some obscure widget or grommet required at short notice by the ship or the scientists. As Chris Adams put it:

> What is not appreciated fully, by scientific participants and the funding bodies, is the extreme trust placed on the ship's agents to act on behalf of the ship and the total reliance upon the agent/Ship's Master relationship. In all the 25 ports at which the ship called over her three year voyage, all the agents involved gave superb service, often in difficult circumstances - or at least, very strange to their normal working routines - and there were very few misunderstandings or omissions.

Adams' trip was eastbound from Barry - initially to London, then via Singapore, Indonesia, Australia, New Guinea, Fiji, New Zealand to Tahiti before returning to his normal place of duty in South Wales. At each stop, Adams had initially to brief the agents on the likely requirements of an unusual, for them, vessel. This covered all the matters referred to earlier, together with any specific points raised by the agents themselves, usually concerning local practices or financial requirements. One of the latter which tends to be taken for granted by those on board RVS ships is that the agent works with the Master to ensure that anyone going ashore is provided with local currency, all of which is accountable. Given that on its circumnavigation, RRS *Charles Darwin* would call at ports as diverse as Muscat (currency Omani Rials), Singapore, Port Darwin and Woods Hole (all different brands of Dollars), and Mombasa (Kenyan Shillings), the Masters had to be not only fiscal mathematicians to work out the appropriate conversion but also to have implicit trust in the agents to provide the appropriate numbers of coins or notes at the right times.

Achieving diplomatic clearance for every cruise also meant that the local British Embassy staff had to be briefed to supplement, or prepare them for, the formal clearance request that they would receive from the Foreign and Commonwealth Office (FCO). NERC and RVS fully appreciated that national perceptions of what the scientific teams were doing could be coloured by political interests, competing international or national funding agencies, or potential impact on local resources. With their local knowledge and contacts, the Embassy staff were in a key position to advise the state authorities of the scientific programmes being supported by RRS *Charles Darwin*, and to allay concerns on these matters. In the event, diplomatic approval for every cruise (bar one) was obtained - albeit with the odd minor hiccough - and was a major contributory factor to the overall success of the voyage.

The other matter that Chris Adams had to check on was the ability of the shipyards in Singapore to undertake a major refit. The comparative term is used here because NERC's erstwhile Director of Scientific Services (Mr Brian Rule) had instituted a cycle of alternate major and minor refits, rather than an annual major refit. In the latter (minor) case, the ship is not normally drydocked, and only essential annual repairs and certification work are carried out; in the former case, the ship is drydocked and a full service carried out, including hull painting. Since NERC requires multiple tendering for a project of the magnitude of a major refit (usually costing several tens of thousand pounds), and since such a refit had not previously been effected in Singapore, Chris made a series of visits to the major shipyards to establish the ground rules for the tendering. Singapore, as the main entrepot of the far east, has little time or berthing facilities for ships which are neither landing nor loading large volumes of income-generating cargo, and at the time (1987-88) was also involved in providing repair facilities for tankers damaged by missile attacks during the conflict between Iran and Iraq. Hence, it was not only necessary to discuss the fiscal niceties with the yards, but also obtain an assessment as to whether or not they would be prepared to service a relative tiddler like RRS *Charles Darwin*. Chapter 7 will show that the negotiations were successful.

Negotiations, it seems, took other forms during this planning trip. Indonesia was an area that NERC ships had never visited, but was a country in which the skills of NERC geologists had long been exercised. The local British Geological Survey resident (and the only bagpipe specialist in Indonesia!) was Dr Sandy MacFarlane, and at the time of Chris Adams' visit his expertise was complemented by that of Professor Mike Audley-Charles, from University College London, and Dr Tim Francis of NERC's Institute of Oceanographic Sciences Deacon Laboratory (IOSDL). To allow RRS *Charles Darwin* to work in Indonesian waters, a formal Memorandum of Understanding was required with the appropriate authority. Chris Adams writes:

> The methodology involved could only be called unique. An early morning rail trip to Bandung to meet Indonesian geologists was an experience never to be forgotten, especially when, on the visit to the Bandung Institute, one realised how fortunate we in NERC were in our working environment - in a vast empty building there were only two working scientists, one working PC, and no funds at all.
>
> The visitors had to sit down and draft a Memorandum of Understanding between NERC and the Indonesian scientists that would satisfy the sensibilities of both the Indonesian and UK authorities involved, and in this delicate exercise Audley-Charles gave invaluable help with his fluent Baharsi. The British Embassy undertook to arrange a detailed series of meetings with various

Government ministries, culminating in a magnificent dinner, and it all provided a valuable lesson in negotiating skills on a basis very different from the conventions normally operating in Europe. Sanity was restored at the end of it all by taking tea with the British Ambassador!

The unusual adventures continued in New Guinea, where Chris was stranded at the airport for three hours because of an oversight by the agent - he had forgotten to organise a car. That period did give him an opportunity to study the local wildlife, principally spiders and small snakes, which prepared his mind for the unusual experience, at least in UK terms, of finding unmentionable reptilian visitors in the shower at the hotel and six (reportedly) functioning elderly Dakotas parked on the grass outside. What it did not prepare him for was the presence of a local man on the floor of the agent's office - on the floor because he had attempted to race a crocodile across a local river, and in coming second had provided the crocodile with a prize of his left leg!

After New Guinea, Fiji seemed likely to be more mundane. However, the Island was still suffering the aftermath of a *coup d'etat*, and on arrival late at night Adams had 'to wake up rather large soldiers who sported open bayonets to be told that a taxi was not on the tourist agenda'. In Fiji, Chris had hoped to organise some co-operative research with the Pacific Rim countries for use of a NERC-developed surveying system, GLORIA, but the political situation prevented their giving financial commitments. The routine tasks of briefing the agent, etc. having been completed, it was time to leave. Adams writes:

> Due to a little local difficulty among the airlines - most of whom had elected to remove Fiji from their schedules - getting off the islands was no easy task. To have the agent put sufficient pressure on an operator to run a plane down to Nadi (the international airport is on the opposite side of the island) in the curfew hours, plus being asked to assist with the aircraft pre-takeoff checks (there were only a pilot, a stewardess and two passengers on a 40 seater aircraft) was impressive. However, worse was to come. The only airline that was still operating was Japanese. So a 747 came in from Hawaii and was rushed by a horde of potential passengers.

It eventually took off for Auckland with over 500 passengers onboard - many standing in the aisles as we lumbered down the runway with total chaos existing onboard.

After these experiences, the relatively straightforward processes in Tahiti, Australia and New Zealand provided a welcome relief for Chris Adams before he returned to his more usual role of ensuring that the operations and logistics ran according to plan. As he notes:

> All the report writing and planning memos, discussions and promises can only be a precursor to the actual event. Much credit must be due to all the ships' agents and staff of sundry foreign organisations and businesses who pulled out all the stops when it was necessary and, in the main, achieved all that was asked of them. The value of face-to-face planning is difficult to quantify, but the knowledge gained from such meetings at least allows the planners and ship's staff to be more aware of likely problems, and allows the people who are being asked to act on the ship's behalf at least to have an inkling of what might occur. In the event the ship went nearly everywhere that the scientific programme required, with full positive approvals granted and logistics achieved.

As the story of the voyage unfolds, we shall see later how relevant these words were.

2
Setting Off

Early in June 1986, only some 18 months after having been accepted into service by NERC, RRS *Charles Darwin* arrived at Falmouth to prepare for the first scientific cruise in what was to become a 39 month voyage before she saw UK shores again. The omens did not look very promising - the instructions to the ship read 'Once alongside, the vessel will shut down to facilitate essential engine repairs' and for the next four days 'Vessel continues repairs', to be followed by 'Vessel goes to dry dock (for a week)'.

But these omens were obviously intended to spur both crew and scientists, for on 23 June (just under a week later than originally planned) the ship sailed in summer weather bound for the Mediterranean and Aegean seas, to conduct research in the latter. The Master, Peter Maw, reported an uneventful passage in good weather, with the ship making 12.4 knots from Falmouth to Patras, in Greece, where the first scientific team embarked.

In the course of her three year circumnavigation, RRS *Charles Darwin* was to be commanded by a number of Masters. Each was a character in his own right - at the time of writing, NERC has yet to appoint a female Master, although a lady Chief Officer has already earned respect from her fellow officers and crew alike for her competence and seamanship. Peter Maw is everyone's idealised picture of a trusty old seadog. Of indeterminate age but certainly not a callow youth, Peter's character is set by the hair and beard which look as though they have been trimmed by an amateur gardener armed with blunt shears. His moustache permanently stained by the smoke from his self-rolled cigarettes and with apparently only a well-worn jumper and randomly creased trousers as formal uniform, Peter nevertheless inspires confidence by his total imperturbability. It was his lot to take charge of RRS *Charles Darwin* on the first cruise of her round-the-world voyage.

A comparison with the start of the voyage of the HMS *Beagle* will show that even 150 years ago, there were differences between planned and actual starting dates. Although Mr Darwin had been told that the ship was due to depart in September 1831, he did not move to Plymouth until 24 October, and HMS *Beagle* did not sail until 27 December because bad weather had prevented it earlier that month. The ship made a landfall at Madeira, but bad weather obliged the captain to sail on to the Canaries. The rumours of cholera in England would have required a 12 day quarantine, which was unacceptable to Captain Fitzroy, and so he set sail again, eventually anchoring in Porto Praya Bay, in the Cape Verde Islands.

The first event of note following RRS *Charles Darwin*'s arrival in Patras was by way of fulfilling a long-standing agreement. The ship's agent was a Mr Morphy (a member of the Morphy-Richards concern) who was also honorary Consul in Patras. Some years before this visit, another of NERC's ships (RRS *Shackleton*) had called at the port, and the Master at the time had noted that the Union Flag flown outside the Consul's office looked a little sad. He had promised that if ever a NERC ship called again he would provide a new flag. Before RRS *Charles Darwin* sailed from Falmouth, this promise was recalled, and on arrival in Patras a new Union Flag was presented to Mr Morphy, who proudly ran it up the flagpole, thus restoring the UK's visible status in his part of Greece.

Our two ships have now begun their voyages, one going South West, the other South East. The departure of each was delayed from that planned, albeit by totally different circumstances. Their paths diverge from their common starting point in southwest England, but the scientists on each would experience similar excitements as new discoveries were revealed, or new insight shed on the mysteries of the deep. Mr Darwin conducted his research with a geological hammer, a sampling net, and an avid interest in sights and creatures and events that were unfamiliar to him. He recorded those finds in letters to his family and his sponsors, and eventually in the book for which he is best remembered. The modern scientist goes looking for data to support, improve or disprove an evolving theory about some (albeit increasingly narrow) aspect of marine science, and his aim is a paper or papers in a scientific journal. Both were moved by the thirst for knowledge, by what Charles Darwin described in the following terms, with which today's scientists would undoubtedly agree -

> ..it appears to me, the doing what *little* one can to encrease (*sic*) the general stock of knowledge is as respectable an object of life as one can in any likelihood pursue.

Was this, then, the spirit in which Professor Michael Brooks, and Dr (now Professor) Michael Collins, both of the University of Wales, had written their proposal for research in the Ionian Sea? If so, that thirst for knowledge became transmuted into a set of scientific objectives which the layman may find difficult to identify. As set out in the ship's *Scientific Sailing Instructions to the Master*, they were:-

(i) To obtain sea surface measurements of temperature and salinity using thermosalinograph whilst on passage

(ii) To study the post-Alpine geological evolution and active tectonics of the western Hellenic arc using single and multi-channel seismic, gravity or magnetic profiling

(iii) To investigate the sediment and water movements in sub-marine valleys and deep water basins.

Before we proceed with the narrative of this particular part of RRS *Charles Darwin*'s circumnavigation, we might perhaps spend a moment or two looking at some of the terms referred to in these chapters, and which will be common to many of the cruises. This cruise covered three of the principal subjects for marine research - sea surface characteristics, the geology/geophysics of the ocean basin, and the movements of the water masses. Each required its own specialised instrumentation, which has been developed over time.

The sea surface in marine research terms is usually considered to be the top 50-100 metres of the ocean, not simply that two-dimensional surface which is visible to the naked eye. To measure the temperature and salinity - the salt concentration - of this surface, a thermo-salinograph is used. This instrument comprises a monitoring system, permanently installed in the ship's laboratory. Seawater is pumped through the meter via a specially designed pump and piping system which avoids contamination of the seawater and minimises any temperature change, and is usually referred to in shorthand terms as the non-toxic system. The monitoring unit records the seawater temperature and salinity, presenting them to the scientist as either an analogue record or a digital signal. An alternative system, when the fixed monitoring system is not installed, is a sensing device which is towed through the water and sends signals to a receiver unit on board ship. The signals provide data of the salinity (usually in terms of conductivity) and temperature of the water mass through which the instrument is being towed. Because of the basic nature of this instrument and the inherent possibilities of its loss or damage in the vagaries of the deep ocean, it is known as an XBT, or Expendable Bathy-Thermo-salinograph; it is a sensor which is deployed on a wire and, as its name implies, is used until water pressure, cable fracture or mechanical damage ends its useful life.

When the scientist is interested in the characteristics of the water column in a specific place, an instrument comprising a recording system known as a Conductivity, Temperature and Depth (CTD) is deployed over the side of the ship. This instrument sends back signals of these three parameters to a recorder on the ship, so that the scientist can deduce changes in conductivity (salinity) and temperature with depth. Of particular interest are sudden changes in either of these, indicating the presence of different water layers. The CTD is more often than not mounted in a rig incorporating sample bottles to collect water from particular depths for analysis on board. Other instruments - such as a fluorometer to measure the light intensity at depth - may also be mounted on the instrument frame. Colour plate 3 shows a CTD with water-sampler bottles attached being deployed over the side of the ship, beneath the surface of the water in conditions that every oceanographer dreams about.

To measure the movement of the water masses, current meters are deployed. These instruments, familiarly referred to as the work-horses of oceanography, are basically the underwater equivalent of the anemometers often seen at ships' mastheads or atop tall buildings measuring wind velocity and direction. The principal differences are that the current meter has to be of much stronger construction for reasons which will become apparent, and it also has to incorporate its own recording system. Current meters are frequently deployed in strings of perhaps five or six. They are anchored to the ocean bottom by a heavy weight, to which the entire string is attached by a remotely-triggered release mechanism. The current meters themselves are arranged on the string at depths selected to cover the scope of interest to the scientists - for instance, in this first cruise, one string carried meters at 1,100 and 1,200 metres above the seabed, whilst others had current meters deployed at logarithmic spacings at 3, 10, 30 and 100 metres above the seabed - where they are suspended by flotation devices such as reinforced glass spheres. The current meter unit is totally self-contained, with a recording mechanism which will record both current speed and direction over periods ranging from a few days to one month or one year, or occasionally even longer, before it is subsequently recovered.

The geophysical characteristics of the sea basin - ie basically the rocks which form the earth's crust under the sea (the lithosphere[4] and asthenosphere[5]) - are measured using seismic equipment. In simple terms, an energy signal (sometimes generated by an airgun, which releases a high energy sound wave, and sometimes by the use of carefully calculated quantities of explosives) is released into the sea. The signal travels through the water, and is reflected from the seabed and the rocks below. The time between the generation of the signal and the reception of its reflection, referred to as TWT or Two-Way Time, is a measure of the depth of the rock strata, with different types of rock giving characteristic signals which can be interpreted by the scientists. The two seismic systems used throughout the voyage of RRS *Charles Darwin* were the single-channel and multi-channel instruments. In the first case, a single hydrophone - a microphone detection system, suitably adapted for use in water - provided a signal for recording on board ship. In the latter case, a large number of hydrophones was mounted in 48 plastic tube modules, which when linked comprised a seismic system some 3 km in length, with suitable spacing and control surfaces along its length. We shall see in later chapters how towing this significant device behind the ship led to some memorable moments.

The geophysical characteristics of the surface of the ocean bottom were ‘viewed’ by sonic means. A system developed by IOSDL used a series of energy-generating devices set (at an angle) in a towed body to provide the sound source, the returning signal being received by the towed body and recorded on board ship. The whole system is known by the acronym GLORIA, for **G**eological **LO**ng **R**ange **I**nclined **A**sdic - and for those whose memories do not go back that far, ASDIC was the acronym for **A**llied **S**ubmarine **D**etection **I**nvestigation **C**ommittee, set up at the end of World War I and perfected in 1920. GLORIA records are processed electronically and photographically to produce the underwater equivalent of an aerial photograph (see plate 17).

Once in the Ionian Sea, the ship and scientists settled into a routine of deploying current meters by day, and seismic profiling - the measurement of the contours of the sedimentary layers below the seabed - by night for the first 10 days. But then, for the second part of the cruise, huge excitement - the routine changed to recovering current meters by day and carrying out grab stations overnight! Grab stations, as the name clearly implies, were predefined locations where the seismic profiling had indicated that there might be rocks on the seabed which could be garnered by a grab lowered from the ship for subsequent geological, mechanical or chemical analysis. In the high-tech world of the ship’s circumnavigation, this is the closest that any of the scientists came to the simple geologising of which Charles Darwin wrote. All this effort was sustained by a high standard of catering, a sentiment which will recur throughout all cruises, and the success of the cruise was celebrated by a special meal at the end. Sadly, details of this menu are left to the imagination.

But this apparently dull routine was not without its moments of, if not excitement, at least note. Two such were recalled by one of the RVS technicians aboard. The presence of an unusual ship in this part of the world excited the interest of the Russians (or more correctly, given the year, the Soviet Union) and for some time the ship was circled by a Big Bear aircraft of the USSR. This in turn caused the NATO allies in the Mediterranean some concern, and the Russian aircraft was shadowed and mildly harassed by two American F1-11 fighters trying to divert its attentions.

The second incident is one that will recur - the apparent lack of basic knowledge of seamanship by some masters of other ships. In this case,

RRS *Charles Darwin* was towing the multi-channel hydrophone just below the sea surface which made manoeuvring difficult. As required by the appropriate nautical regulations, the ship was flying the necessary signals (black balls during the day, the relevant lights at night) at the masthead warning other ships of this constraint to manoeuvring. One small vessel appeared not to see these signals, and did not respond to attempts at contact by VHF radio or Aldis lamp. In the end, the small ship's course took her relatively close to the stern of RRS *Charles Darwin*, and the only way that a scientific catastrophe was averted was by setting the controls on the hydrophone to make it dive to more than 30 metres below the surface so that the errant vessel passed safely over it.

Data collected from this initial cruise are being analysed presently by the UK Principal Scientists (Professors Brooks and Collins), together with their Greek collaborators in Patras and Athens. Indeed, preliminary interpretations of the data sets (Pattiaratchi and Collins, 1988) have been integrated already into the European Union-funded programmes concerned with water and sediment movement in Greek seas.

Patterns of surface water movement appear to be controlled by meteorological forcing, whilst a tidal signature (even for the Mediterranean Sea!) is evident in near-bed current meter records collected even in the deeper water areas.

The sedimentology and biostratigraphy of sediments within the piston cores collected from the sea bed, in water depths ranging from 200 m to 3,400 m, reveal considerable displacement of the deposits. Such characteristics represent the deposition of sediments in a tectonically-active region of the eastern Mediterranean Sea (Burrows, 1994). On-going investigations are concerned with an inter-comparison of mechanisms generating currents in submarine canyons (Pattiaratchi *et al*, in prep), and a collation and interpretation of sedimentological and oceanographical data from the Hellenic Trench, eastern Ionian Sea.

And so the first cruise ended when RRS *Charles Darwin* steamed again into Patras on 17 July, disembarked the first scientific team, embarked the second team, and refuelled etc. She then sailed onwards towards the Suez Canal, the Red Sea and the Indian Ocean. But before we leave Patras, where the ship actually stayed for a day or two, we must introduce the first of several personal stories germane to the voyage.

The Chief Engineer on this first cruise - as on several others throughout the circumnavigation - was George Batten. A man of enquiring mind,

George thought that it would be a good idea to use the opportunity of the stop in Patras to visit the temple of Delphi, but first he had to supervise two days of taking on bunkers (fuel, to the uninitiated) after which he had, to use his term, 'earned a day off'. A small party, comprising Jeff Baker (the Radio Officer), Phil Evans (the Second Mate), Kay Potter (the RVS Software specialist) and George, set off from Patras by bus to catch the ferry to Navpaktos and then another bus to Delphi - a total distance of some 50-60 miles. Delphi excited comments of sheer delight from George, and some of these were shared by Kay.

On return to Navpaktos, the party decided to enjoy dinner and a drink before catching the ferry back, and this delightful interlude extended well into the evening. When they felt it was time to return to the ship, they discovered that not only had the ferry stopped operating for the night, but it had done so because the last bus back to Patras had left some time previously. The rapport between George and Kay had, by this time, developed to the stage where they proposed that all four in the party spend the night on the beach to await the morning's ferry. Phil and Jeff were less than enthused by this idea, and persuaded a local stall-holder to ferry all of them across to the Patras side of the water, where they were able to find a taxi back to the ship. The story continues in the Seychelles!

In Patras, there was a change of Master, and Keith Avery took over from Peter Maw. At the time that RRS *Charles Darwin* was making this particular cruise, NERC had a policy of having only three substantive Masters, and appointing Chief Officers to Masters' posts on a rotating basis; Keith was one of the Chief Officers so appointed (although he was promoted substantive Master in 1992). In appearance, Keith was the direct opposite to Peter Maw - clean-shaven, non-smoking, always wearing a sparkling white shirt, trousers with proper creases and well-shined shoes - but enjoying the same sort of respect from his colleagues and the scientists.

On the ship's arrival at the start of the Suez Canal, when anchored in the Husein Basin, the Master had to institute security procedures rarely necessary in Barry. The ship had to be kept physically secure, with only one door openable from the deck, and the constant presence of a member of the crew or RVS support staff on watch. The passage through the Suez Canal was uneventful, RRS *Charles Darwin* being sandwiched between two Russian vessels at the end of a convoy of 16 ships.

During this period of time, NERC had hopes of further commissioned research in the Middle East, and hence special efforts were made to convey a good impression. When the passage through the Red Sea was being planned, it was decided to make a courtesy call at the Islamic port of Jeddah, and hold a reception on board ship. Following discussions with the British Embassy in Riyadh and the Consulate General in Jeddah, plans and guest lists were finalised. In this process, the King's office in Riyadh requested that three Saudi observers sail from Suez to Jeddah, and subsequently Dr Moamar (vice-Dean of the Marine Facility at King Abdul Aziz University, Jeddah), Mr Al-Ghamdi (Director of Natural Resources at the Marine Environmental Protection Agency) and Lieutenant Shamiran (from the Military Survey, Riyadh) were duly made welcome and given a hectic introduction to the ship, its staff and systems. In return, the observers gave ship's staff valuable insights into the culture of Saudi Arabia.

But it was not only education that was exchanged. In the relatively short spell for which the observers were on board, a mutual liking and respect developed between them and the ship's staff, culminating in an exchange of gifts. Mr Al-Ghamdi presented the ship with a four-volume *Natural History of Saudi Arabia*, Dr Moamar presented some pictures of Arabian scenes, and the three jointly presented the ship with a typical Arabian silver vessel for holding rosewater. In return the Master (Keith Avery) presented each of the observers with a piece of Wedgwood (Charles Darwin having married Emma Wedgwood in 1839). Keith had been aware, before he joined RRS *Charles Darwin* for this cruise, that some ceremonies of this nature might be required and had had the foresight to visit his local antique shop in Tintern to acquire (on a 'sale or return' basis) the three pieces - one dated 1809 (the year the great naturalist was born), one 1839 (the year he married), and one 1882 (the year of his death). Of course, it would be possible to argue that such a gesture was not strictly necessary, and the ship could have responded to the generosity of the observers with a simple note of thanks, but the fact that at least one of the Wedgwood pieces was later observed to have been given pride of place in the new owner's reception area suggests that the value to NERC (and, by extension, to UK) interests far outweighed the investment that the Master had made.

Early on the morning of Monday 28 July, RRS *Charles Darwin* berthed at the Saudi Arabian port of Jeddah for a courtesy call. The courtesies began immediately, with guests from the Marine and Environmental Protection Agency being shown around the ship. The afternoon was devoted to spit and polish preparatory to a formal reception in the evening, although the officers and scientists left the ship for a visit to the King Khaled University. A measure of the importance attaching to the trip was shown by the fact that Dr Moamar had recalled a number of staff from holiday in order to show the UK visitors over the University's facilities which have been described as being like the well-known hymn-book *Ancient and Modern*. However, the Marine Department of the University operated a number of small craft, more akin to yachts than to RRS *Charles Darwin*, and it was felt that the exchange of views had been of mutual benefit.

The evening reception was judged a success by the ship's staff, but only time and others will be able to give that judgement any objectivity. One unusual fact that stays in the Master's memory is that it was the only occasion on which he savoured the delights of cactus fruit, which he said was 'quite tasty'. A chance remark by the Pilot suggested that the arrangements for the reception were a dress rehearsal for a visit by

Plate 5: Jedda reception from LMS

the Royal Yacht *Britannia* later in the year! Whatever history records about the event, it was marked by yet another exchange of gifts, this time a small silver salver from the guests being reciprocated by a shield bearing the ship's crest.

Two points of particular interest here are that firstly, the Principal Scientist for the cruise that would start from Jeddah was a lady - Dr Carol Williams - who had to take special care not to antagonise the male-dominated ethos of the host country. The second was that the reception on board had to be timed to avoid the evening hour of prayer. Dr Leonard Skinner, at that time Head of Service at RVS, and Dr Keith Harrap, Head of NERC Marketing, both attended to support the Master and Dr Williams. A photograph of the Principal of the University of Jeddah being welcomed aboard RRS *Charles Darwin* is shown above. One observer at the reception commented on the 'wide range of drinks that could be prepared from fruit juice'!

NERC had produced some display boards for use during this and later events, but two of them were delayed and had to be flown out at the last minute. These boards comprised photographs and promotional material on a rigid plastic sheet, 1.5m x 1m x 1 cm, and were shipped to Jeddah via Cairo. Because of administrative problems at Cairo Customs, the boards either went astray or were lost, and it is interesting to speculate that there may now be a modest Arab house whose roof displays, either to its inhabitants or to any passing birds, a picturesque summary of NERC research capabilities in particular areas of environmental science. In spite of the loss of these two boards, the display material on the ship impressed journalists sufficiently for the local newspapers to carry reports that the exhibition

> ..gave information on the use of polymers to aid the planting of trees in the deserts, outlined research to monitor and tackle atmospheric pollution, and gave details of NERC's new hydra-equipment which assesses how much water has evaporated from a particular area. (*Saudi Gazette*, 28 July 1986).

The cruise that Carol Williams had planned was a seafloor topographical survey of the Red Sea prior to a possible visit by the Ocean Drilling Program (ODP) ship, the *Joides Resolution*. ODP is an international exercise, a continuation of similar efforts which had been carried out for some time, to drill deeply into the earth's sub-sea surface for material on which to develop or expand theories about the evolution of the rocks and their movements. It used techniques based on, and developed from, drilling methods evolved for the exploration and exploitation of oil reserves from beneath the seas and oceans. Carol Williams' survey depended upon obtaining permission (clearance) to work in Saudi Arabian, Sudanese and Ethiopian waters. To quote the official cruise report

> ...In the event, permission was granted only to work in Sudanese waters of less than 1,000m depth. No response was received from the Ethiopian government, while Saudi Arabia and the Saudi-Sudanese Commission for Red Sea Resources refused permission in their waters. Since all the high priority drill sites were in the central rift area, the original objectives for the cruise had to be abandoned.

RRS *Charles Darwin* was, of course, on passage through the Red Sea and thus offered the scientific team on board a valuable, if not unique, opportunity to carry out some investigations, even if not that which they had planned. Dr Williams rapidly sketched out a new schedule, to map the poorly known older series of magnetic anomalies in the Somali abyssal plain. This flexibility highlights two points which recur throughout the circumnavigation. The first is that the ship is an expensive resource which happens to be at a particular point in the oceans at a particular time: if it is not used to gather data from that location, the opportunity may not occur again for perhaps a very long time, perhaps never. The second point is that the Principal Scientist needs to be aware of the value of this resource to him- or herself, and if circumstances negate original plans, an alternative with some scientific value must be produced.

The history of this cruise is one of triumph, tragedy - and a large element of what, in retrospect, can only be termed comedy. Firstly, the triumph - in spite of only being able to work in Sudanese waters of less than 1,000m, gas blow-out fissures were observed and it was therefore probable that the ODP drilling would be precluded on safety grounds.

The tragedy and the comedy begin by being linked by the weather. All was well in the Red Sea and the Gulf of Aden, with only the high sea temperatures to concern the ship's engineers (of which more anon), but once around the Horn of Africa (officially, Cape Guardafui) the weather changed and the ship headed into a monsoon with a windspeed measuring 8-9 on the Beaufort scale. This caused the average passage to drop from about 290 nautical miles (nm) per day in the Red Sea to less than 60 nm at the height of the storm, so that the planned scientific work dropped behind schedule. The weather also introduced its own element of pure comedy - one part of the cruise was concerned with gathering dust particles for later analysis, but the carefully rigged filter was washed away by the monsoon!

However, the pure comedy can, as so often, be laid at the door of administrative error. RRS *Charles Darwin* was on passage from Jeddah to Port Victoria in the Seychelles, and the scientists had been expecting to be able to use their gravity meters and magnetometers up to the moment of arrival. When approaching Seychellois waters, the ship was advised by RVS that no permission to work in those waters had been received, and so it was decided to perform a CTD drop (not normally part of a geophysical cruise) just outside the Seychelles' territorial waters. Just after completion of the CTD drop, a telex from RVS, sent during the previous cruise, was found in one of the ship's files giving permission to work in Seychellois waters! It was then too late to do more than decide upon the minimum survey leg possible to arrive at Port Victoria on time.

And so RRS *Charles Darwin* arrived in the Seychellois port of Victoria to complete her first scientific cruise 'east of Suez' and to embark a new team for the first of the many cruises that would take the vessel all around the Indian Ocean for the next year. But for two of those leaving the ship, the Seychelles marked more than the end of a cruise.

Let George Batten tell it in his own words.

> The ship arrived in Victoria on 10 August, and because of the logistics, return flights to the UK for the ten people leaving the ship were not available until 16 August. Kay and I decided that we would make the most of this period, and we hired a Minimoke to see the island. As time passed, our friendship became deeper, and

neither of us fancied returning home. We therefore asked Air Seychelles whether there were seats available on the flight a week later than the one on which we were booked, but their advice was that they could not confirm a revised booking before the planned flight had departed. We decided to take the risk, and advised RVS that we would not be returning with the others. This in itself caused a flutter in the dovecotes, because the (then) Marine Personnel Officer said that the tickets had been booked at a discount rate, and changing them would mean a financial penalty for RVS: this advice was later shown to be erroneous. Our next step was to visit the Seychelles Tourist Office, and ask to extend our visa for a further week. It seemed that the official there was prescient, because he gave us a two-week extension, and in the event we decided to stay the extra week, and arrived back in UK two weeks after the remainder of those who had left the ship at the same time as we.

George Batten and Kay Potter were married in August 1988, and now have a son, Sam, born in 1993.

---

[4] Lithosphere - the more rigid outer layer of the earth overlying the asthenosphere

[5] Asthenosphere - the upper mantle layer from about 65 to 165 km depth

3
The Indian Ocean

Among the proposals for research in the Indian Ocean had been one from a group of marine chemists. Sometimes considered the poor relation of marine science, chemistry through the 1970s and 1980s was beginning to become a subject of significance with the development of techniques that provided accurate analysis of trace elements in seawater. These led the scientists themselves towards a better understanding of geochemical activities such as the transformation of trace elements to the deep ocean through biogeochemical processes. Some research had already been effected in the Atlantic and Pacific Oceans, but, to quote Professor Dennis Burton, one of the co-Principal Scientists on CD15/86, 'the Indian Ocean represented a major gap in the global picture'.

The two terms geochemical and biogeochemical require some explanation. The prefix 'geo' means earth, so formally, geochemistry means a study of the chemistry of the earth. In the marine environment, however, this chemistry involves not only the study of the elements and compounds that exist in the sea and on the seabed, but also the means by which these may change in chemical form during, for example, their passage from atmosphere to sea-surface to seabed. A biological component of study is introduced by the effect of the planktonic and higher order food chain. So for clarity, this branch of geochemistry is referred to as biogeochemistry, and this has become a discipline in its own right (*see Chapter 4*). However, the boundary between the two disciplines is unclear, and this record accepts each team's definition of its own area of study without comment.

Because no one university had a Department of Marine Chemistry of sufficient size to support a cruise solely from its own resources, a proposal involving several universities was put forward and received NERC support. This had the benefit that it brought together on RRS *Charles Darwin* researchers with interest and expertise in particular constituents, and added a richness to the study. As an indication of the breadth of interest in the topic, the cruise was led jointly by Dr Harry Elderfield (Cambridge) and Dr (now Professor) Dennis Burton (Southampton), with support from the University of Liverpool, University College North Wales, and the Massachusetts Institute of Technology. RVS also provided a support team, as usual, but in addition it had made a major contribution by the development and production of an ultra-clean containerised laboratory to permit the sensitive chemical analyses that the chemists required, the first time that such a containerised clean laboratory had been used on a UK research vessel.

The formal aim of this cruise was

> ..to evaluate the geochemical cycling of trace metals in oceanic waters by means of a longitudinal chemical section in the western Indian Ocean.

but this is perhaps more understandably set out in the opening sentence of one of the papers resulting from the cruise (Morley, Statham and Burton, 1993) which states

> The increasing database of accurate measurements of trace metals in oceanic waters has provided a coherent basis for the interpretation of variations in concentrations within the surface and sub-surface waters of the world oceans in terms of the interaction of geochemical transfer processes and the circulation and mixing of water bodies. In terms of global coverage, important gaps have remained: information is notably lacking for the source regions of intermediate and deep waters and, of importance from the standpoint of variability between the major ocean basins, for the Indian Ocean.

In other words, the availability of accurate measurements of trace metals allows not only an assessment of the variability of these within and between water masses, but in some cases could provide a means for assessing the origins and movements of the water masses themselves. Morley and his co-workers used the data gathered on this cruise to conclude, *inter alia*, 'that deep water concentrations of the recycled elements are intermediate between those for the North Atlantic and North Pacific Oceans, as would be expected from known patterns of deep-ocean circulation', thus confirming the second of these hypotheses. Also from the same cruise, Bertram and Elderfield (1993) show that 'inter-ocean variations in neodymium isotopic compositions and Nd concentrations cannot be reconciled unless particle-water exchange is invoked'. In simple terms this suggests that some interchange between particulate and dissolved neodymium (Nd, one of the Rare Earth Elements) in the deep ocean is needed to explain the measured results.

But not all of a cruise involves such esotericism - even scientists have human failings. The Master, Geoff Long, in his report on this cruise notes that all went well, except for two events. The first was when one of the scientific team tried to make cheese on toast in the vertical toaster. The second was when a scientist put his clothes and a quantity of soap powder in the spin drier, switched it on and left it: the result, to quote, 'was that the laundry looked as if it had been filled with HiEx foam[6]'. Such experiments should merit an exclamation mark, but since they were not so accorded by the Master, the writer is loath to improve upon his prose style.

Geoff Long had taken over from Keith Avery in Victoria, and was another of NERC's long-serving, not to say long-suffering, Chief Officers who were called upon on a regular basis to act as Master on any of the Royal Research Ships. Closer in appearance to Peter Maw than Keith Avery, Geoff's approach is also closer to the pragmatism of the former than the formality of the latter.

Whilst this first cruise started and finished in the Seychelles, the bulk of the cruises reported in this chapter and the next were based upon the port of Mina Qaboos, now renamed Port Sultan Qaboos, in the Sultanate of Oman. Hence a few words about the country are in order. The Sultanate of Oman and the United Kingdom have ties of friendship going back over many years. The present Sultan is an Anglophile, having been through the Royal Military Academy Sandhurst, and maintaining a home in UK. On the other side of the coin, at the time of the ship's visit several of the senior technical posts in Oman were occupied by ex-patriate Englishmen. That said, it was not until the present Sultan replaced his father in 1970 that efforts to modernise the country began in earnest, and at the time of the visits by RRS *Charles Darwin*, the country remained virtually closed to tourism.

One of the most visible aspects of this policy was that potential visitors to Oman required *No Objection Certificates* (NOC) from the Omani Embassy in their home country before they would even be allowed to board the flight to Oman. Obtaining such NOC took time and effort, but clearly support for the scientific activities of the ship could not proceed without the attendance of both scientists and RVS staff. In the majority of cases, all went well. However, the experience of one of the early parties raises a wry smile. Ken Robertson, of RVS, recalls:

> A group of us travelled to Heathrow to board the evening Gulf Air flight to Muscat. At the check-in desk we were asked for our NOC. These had always been arranged through the agent who regularly met us at the airport in Muscat and led us through the formalities. On this occasion the RVS request to the agent for NOC had been rather late but we were informed in Barry that everything was in hand and would be as normal.
>
> Having explained this, we were allowed to board the aircraft and off we went! Arrival in Muscat was normal until we presented ourselves at the appropriate desk before immigration and were told there were no NOC in our names. I believe this was on a Thursday and the Omani officials said that the appropriate Government department did not reopen until Monday of the following week. Without NOC we could not be allowed into the country. Everyone was most polite and courteous but rules are rules!
>
> With profuse apologies our passports and baggage were confiscated and we were told that we would have to remain in the embarkation area of the airport until a solution was found. Furthermore, the laws said that we could only stay there for a

maximum of 24 hours and if a solution was not then apparent we would have to return whence we had come (ie to the UK). The confusion and annoyance can be imagined and we did our best by contacting the agents by telephone but with no immediate results. They have to live by the same rules after all!

The airport got quieter and quieter and we eventually found the departure lounge with a small restaurant attached. The bench seats looked like the most comfortable place to spend the night, so we stretched out and tried to get some sleep. Luckily nobody in the airport objected. During the night a representative from British Airways came to see us since he had learnt of our plight. He confirmed that if the worst came to the worst, he would guarantee us enough seats on the aircraft coming through from Australia to London the following day. Not much consolation but better than nothing. Sleeping in an airport lounge is never the most comfortable thing to choose but there was no choice. I recall that the loo paper ran out in both the available toilets but at least we were not cold.

The following morning we were offered a very good breakfast at the expense of the airport as soon as the restaurant opened and things began to seem less disastrous. Within an hour we learnt that the agents had worked some magic and all but one of our party would be allowed into Oman. John Price was the unfortunate exception and because of an administrative mix-up between his name and that of Brian Price, who had sailed on an earlier cruise, there was no chance of procuring an NOC before the maximum 24 hours delay had elapsed. John duly boarded a flight back to UK, only to return to Muscat a few days later when the formalities had been completed.

Once we were re-united with our passports and baggage and with apologies again from the staff who were only doing their jobs, the agent took us to the hotel. An explosive telephone call to Barry made us feel vindicated and the rest of the portcall went reasonably smoothly.

A postscript to this story appeared in *The Beagle Times*, a shipboard newspaper which, among serious summaries of the UK news and sports results, carried a 'Rumour has it...' page. In the edition of 1 March 1987, one item in this page read..

> ... beneath his abundant black beard, John Price is that well-known wanted criminal, Lord Lucan, hence the reason that the Omanis didn't let him into the country.

Once docked in Mina Qaboos, Geoff Long handed the ship over to Captain Philip Warne. Phil Warne was the most senior of the three permanent Masters, and was probably the most familiar with the ship as he had stood by when she was under construction at Appledore Shipbuilders. Tall, aquiline in appearance, superficially aloof on first acquaintance, Captain Warne conveyed an impression that his true destiny should have been on the bridge of a majestic ocean liner, rather than in command of this little 70 metre vessel, an impression re-inforced by the fact that almost alone among NERC's Captains, he more often than not routinely wore his cap. Assured of his own competence, and secure in the knowledge that retirement was measurably close, Phil had a nice turn of phrase which he was not afraid to use when occasion required.

The first cruise out of Oman was concerned with palaeo-oceanography, the study of the chemistry involved in the transport and transformation of sediment from the sea-surface to the seabed. This became, in the formal language of the Sailing Instructions:

> .. to study the chemistry of the sediments, pore waters and overlying waters in the area of the north-west Indian Ocean where the oxygen minimum zone in the water is well developed. The strategy of the cruise is to sample both waters and sediments at depths above, within and below the oxygen minimum zone (22-1,200m). ... A secondary objective will be to investigate change in sedimentary facies[7] and in the physical properties of the sediments across this depth zonation.

Put more simply, when material from either the atmosphere or other sources is deposited on the sea surface, it can be expected to sink to the seabed at a rate which is, *inter alia*, a function of its particle size.

However, much of the finer material will be used by the microscopic life in the sea - the phytoplankton[8] and zooplankton - as a food source, which will then either contribute to the growth of these animals or continue its way to the seabed as waste products from them. In both cases, the original form of the deposits will be changed, and may also be changed by subsequent absorption into, or passage through, higher forms of life in the sea before reaching the seabed. In all cases, the constitution of the matter on its way through the water column will give positive indication of its route, period of residence, and age to the scientist. For instance, an element with a particular (and measurable) isotopic composition may be absorbed into the shell structure of the minute animals classified as coccolithospores; when these animals die, the isotope in their shells will show when that element was absorbed, and give a clear indication of how long it has been in the water column. Similarly, the collection of coccoliths from the sediment of the seabed will indicate not only the rate at which such material is being deposited but also its age. The proportion of material of similar chemical composition which has been deposited without passing through the biological chain compared with that which has will also yield important information about age and deposition rates.

Once on the seabed, the deposited material will build layers, which although they may not be as uniform as, say, rows of bricks in a building, will provide data about the age of the deposits - hence the palaeo (paleo in American), meaning old or ancient, in the title of the discipline. The chemical or geological history revealed by these layers will show how climate has changed with time by, for instance, indicating the timing and influence of ice ages.

In the 1960s and 1970s, the theory of plate tectonics gave a new impetus to what Charles Darwin had termed geologising: it was, however, no longer possible to study the subject in sufficient depth with only a geological hammer! A global overview of the effect of tectonic plates is shown in Colour Plate 2. To measure the actual movements of the plates that make up the surface of the globe requires techniques undreamed of in the naturalist's day. The two cruises reported now will expand on these.

Victoria A Kaharl [9] writes

> ..The geophysicists are the poets of oceanography. Their lot is not to pry off chunks of earth but to measure the invisible and to infer

characteristics, eventually to arrive at some elusive truth of what is happening beneath the seafloor. If they can learn what happened a year or a million years before, they can better predict what will happen in the future. Their measurements of magnetism, gravity, electrical conductivity and the propagation of sound waves through rock speak the language of squiggles, blotches, and stripes, which they translate into characteristics of the planet - how elastic, porous and dense it is.

More prosaically, the formally defined aim of the first geophysics cruise out of Oman was:

.. to study the structure and sedimentology in the Gulf of Oman, north west Indian Ocean ...by seismic profiling and wide angle seismic line to seabed receivers and land receivers north of Pasni, by taking seabed and sediment cores across the continental margin.

Dr (now Professor and an FRS) Bob White, of Cambridge University, used seismics as his principal tool. For one source of energy, the Cruise Report (White, 1987) says simply that '..(explosive) charges of 25kg, 100kg, or 200kg (were) fired at 5 minute intervals'. For comparison, a recent history of terrorist bombs reveals that less than 0.5kg can totally destroy a car and cause considerable structural damage in the immediate area, and the amount of explosive used to severely damage the Grand Hotel, Brighton in 1984 was only 11kg[10]. What this does illustrate quite graphically is the immense amount of energy needed to propagate a signal up to 100 kilometres through the rocks to reach deep below the seabed. The refractions through and reflections from the rock strata generated by these explosions were measured by a number of Ocean Bottom Seismometers (OBS) developed by the University of Cambridge.

The potentially damaging effect of such large quantities of explosive is well recognised by both the scientists and the RVS technicians who handle and deploy it. The technicians, particularly, receive formal training as 'Shot firers' and a certified shot-firer is always in charge of handling the explosives and associated items, such as detonators, on board the ship. However, regulations exist to govern the handling, storage and use of the material, and all involved in the actual research applications are professionally advised before they are allowed anywhere near the explosives and detonators necessary for the seismic surveys. And one final protocol is always carefully observed - when such surveys are being planned, not only does the clearance procedure involve diplomats, it also stretches to the world's navies who may be operating in or near the waters where the survey is to take place: large quantities of explosives going off underwater could ruin a submariner's day! At the other extreme, one of the RVS technicians involved in the deployment of the explosives speaks with almost awe of the effect of an explosion on the unruffled surface of the water at night. The ripples can be seen to spread from the source, moving the photo-plankton that give the ocean its night-time luminescence into a circular pattern, with brighter pools of light where there is interaction between the ripples and the water flow caused by the passage of the ship and the hydrophone array.

The recording system used was the Multi-Channel Digital Seismic Acquisition System (MCDSAS). For this, a 3 km 'streamer' is towed behind the ship, and kept at a preset depth by control surfaces ('birds') along its length. Within this streamer are some 48 separate detector sections, each containing a large number of sensitive microphones (or more strictly, since they operate in water, hydrophones) which detect the reflected energy from the rocks beneath the ocean bottom. Because the streamer is being towed at a set speed by the ship, the hydro-phones will detect the signals from a series of energy sources displaced in time and space by predictable amounts, and signals returning from a particular feature of the subsurface will be detected at different times by different hydrophones within the streamer. All these data are recorded digitally, and subsequent sophisticated computer analysis provides a highly detailed impression of the subsurface geophysics of the area being surveyed from which the scientists can deduce much about its structure, history and evolution.

Because of problems in obtaining clearance to work in the waters off Pakistan - genuine, this time, not due to a misplaced telex!- the actual area of research had to be altered at the last minute. But it was not really last minute - Bob White notes in his cruise report

> If it hadn't been for the fact that I had already thought about an alternative programme outside Pakistani waters, that this area is almost virgin territory with outstanding geological problems still unsolved, and that the Omani authorities were so quick to give

clearance, then this cruise would have been a huge disaster.

Marine science may well transcend international borders, as Sir Anthony Laughton notes in his foreword, but the facts are that there are national sensitivities that may affect the best laid plans of mice and men, and Bob White's contingency planning proved its worth. Of course, two factors may have contributed to the readiness of the Omani authorities to give clearance for research in their waters - the fact that RRS *Charles Darwin* was essentially based in Oman, at the port of Mina Qaboos, for some months, and that, before his cruise, Bob White had given an informal talk to a group of commercial and governmental officials outlining his scientific objectives, and the nature of the seafloor in the Gulf of Oman from his previous research there.

That said, the conditions for geophysical research were ideal - warm

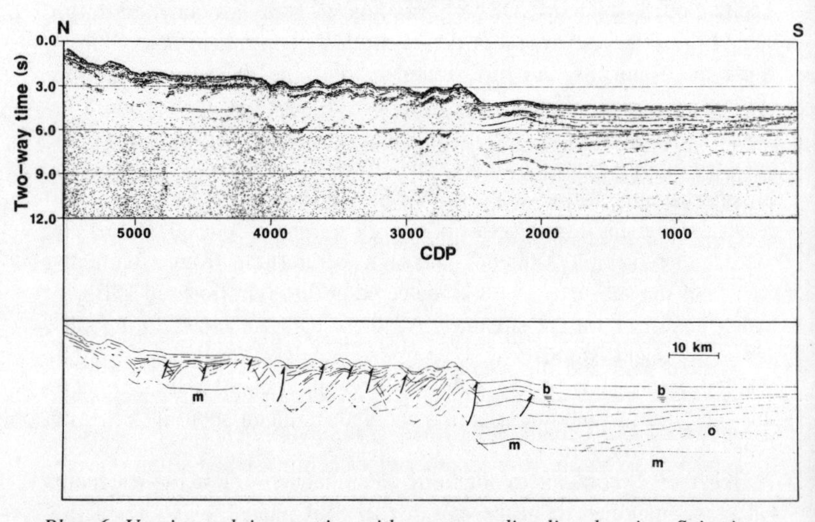

Plate 6: Unmigrated time section with corresponding line drawing. Seismic data were collected using a 48-channel, 2.4 km hydrophone streamer and a 25L (1,500-in^3) tuned air gun array, both towed at 10m depth. Recording system was a Sercel SN358 operated at 4ms sample rate. Standard processing sequence includes predictive deconvolution, normal moveout correction, and stack with time and space variant bandpass filters. A mild automatic gain control (1s window) is used for display. Bold lines show interpreted thrust faults, ***o*** *is oceanic basement,* ***m*** *is the water bottom multiple and* ***b*** *denotes seismic bright spots. Vertical exaggeration is 3.5 at the sea bed.*

weather and calm seas. Too warm, sadly, for the ship's engines where flow restrictions (caused by plankton bloom in the temperatures of the Arabian Gulf) in the Low Temperature Heat Exchanger (LTHE) coupled with relatively high sea temperatures necessitated a 24 hour stoppage to strip down and clean the LTHE. Sadly, such a procedure was necessary because RRS *Charles Darwin* had been built to a price rather than to a particular specification, and the price had not stretched to a second LTHE. For her normal areas of operation - primarily the Atlantic Ocean - such a constraint was unlikely to have any serious impact upon science, since plankton bloom is far from common in the cool-to-temperate waters of the Atlantic.

The LTHE comprises basically a set of titanium plates between the fresh water of the engine systems and the salt water of the sea, and these plates provide an interface for the cooling of the fresh-water needed to keep the auxiliary equipment (such as the air-conditioning plant) cool. What happened in the Indian Ocean was that plankton adhered to and grew on the outer (seaward) side of the titanium plates and restricted the flow of the seawater. Without adequate cooling, the ship's services could not function properly, and so the Engineers had to close them down, remove the titanium plates, scrub them clean and replace them.

Once this problem was rectified, and a decision taken to carry out the science in a less politically sensitive area, Dr White was able to report

> ..the science was outstandingly successful. The equipment worked well and the weather was near perfect. We have some excellent data from Omani waters, and it is a tribute to the speed at which the Omani authorities granted us clearance that we achieved so much from an otherwise seemingly bad situation..

But this apparent equanimity hides an interesting comment from the ship's Master in his report on the cruise. The first notes '... There was a setback at 1800 (on 11th December 1986) when a shark attacked one of the array sections and cut off the records from the rear two thirds (of the seismic array)'. Such problems are unlikely to affect the majority of geologists or geophysicists, and may be thought to add to the interest of marine science! However, for some reason, the hydrophone array being towed behind the ship at speeds from 4 to 6 knots (depending on a variety of factors from the weather to the depth at which the hydrophone was to be flown) seemed to be particularly attractive to sharks, and the loss of one or more sections due to these attacks was far from uncommon. One explanation offered for this attraction is that the control sections (or birds) cause water turbulence when their fins move to keep the array at its specified depth, and this turbulence stimulates interest from the shark, who sees this sinuous body moving slowly through the water. All that the shark gets for his interest is likely to be a mouthful of the oil that fills the hydrophones to keep them neutrally buoyant and the possible loss of some teeth. What the scientific team on board lose ranges from a period of time while the array is recovered and repaired through to a significant reduction in the number of data channels that can function properly.

Some of the group of research scientists took advantage of this rare opportunity of travel in Oman to visit the spectacular geology exposed there. Over a huge area of northeast Oman a large segment of former oceanic crust was forced up onto the land about 120 million years ago, forming the Semail ophiolite. It has become one of the classic areas to study oceanic crustal structure in the comfort of dry land.

Bob White particularly recalls the quiet solitude of the Semail mountains in stark contrast to the ceaseless throbbing of the ship's diesel engines and the continual 24-hour activity aboard ship. During the trip which he arranged following the end of the cruise, a party of scientists was able to visit the Semail to see the copper deposits which had been mined extensively at the time of the Phoenicians. He found it a strange experience, walking around and touching a section of seafloor which in all probability was just like the crust thousands of metres beneath the ship and which they had just spent a month studying.

Before reporting on the next cruise, which began in Port Victoria in the Seychelles, it is worth noting that RRS *Charles Darwin* attracted media attention wherever she went. The *Seychelles Nation*, in a full-page article of 4 September 1986, set out most of the details of the vessel's cruise plans for the Indian Ocean. With reference to the geophysical cruises, the *Seychelles Nation* notes that..

By examining the records of these (seismic) signals the marine geophysicists will be able to study the tectonics of a massive sedimentary wedge as it is scraped off the undermoving Arabian oceanic plate where it approaches and meets the Asian continent. The Makran off Pakistan and Iran is one of the best known examples of an accretionary sediment prism at a convergent plate boundary and this investigation is expected to yield a fascinating insight into the fundamental geological processes at work beneath the sea bed.

The third geophysical cruise in this area of the Indian Ocean was led by Dr Alistair Baxter then at the City of London Polytechnic (now at the University of Greenwich). The formal remit of the cruise was to:

> undertake petrological, geophysical and sedimentological studies of the Rodrigues Ridge (18° 30'S to 19° 40S and 58°E to 64°E) and a site survey of four proposed ODP drilling sites on the Mascarene Plateau.

As spelt out by the *Seychelles Nation*, this became

> ...a 22 day research cruise to investigate the geological relationship of the Mascarene Islands (Mauritius, Reunion and Rodrigues) with adjoining submarine structures which include the Rodrigues Ridge, the Mauritius Trench and the Mascarene Plate. By coring the seabed, dredging up samples, and employing seismic profiling, the researchers hope to provide important new information on the regional structure, tectonics and volcanism of the western Indian Ocean.

All of this verbiage reinforces an increasingly significant aspect of the circumnavigatory cruise - that the ship and her scientific equipment comprise an expensive resource which happens to fill a hole in a particular part of the ocean at a particular point in time. It may be that the geographical point is important scientifically (as, for example, in the case of studying a specific geophysical phenomenon) or that the timing is significant (as in the case of cruises reported later where the effect of the monsoon was to be studied). Because of this, marine scientists in general try very hard to co-operate in the exploitation of this resource in order to obtain the maximum scientific data from the presence of this resource.

Certainly, in the case of Alastair Baxter's cruise, there were American scientists on board who wished to maximise the value of the passage from Oman to the working area to carry out basic surveys to assess the suitability of the area for later drilling by the ODP vessel *Joides Resolution*. Leg 115 of ODP had two primary objectives:

1 To sample the volcanic basement rocks of the Mascarene Plateau - an objective not previously achieved by either dredging or drilling - in order to investigate hotspot associated volcanic lineaments in the Indian Ocean

2 To sample the sediments at various depths within a limited equatorial region.

These surveys used the routine tools of geophysical work - gravity and magnetic measurements, 3.5 kHz and 10 kHz Precision Echo Sounder bathymetry, coupled with seismic surveying using airgun sources and single channel and multi-channel hydrophone systems. A word or two about two of these techniques might not come amiss at this point. The first of these - gravity - is essentially a function of the geoid - the shape of the globe. One of the basic facts of schoolboy physics and applied mathematics is that the geoid is not a perfect sphere but an oblate spheroid - ie flattened at the poles. If one weighs a given mass using a sensitive spring balance, the perceived weight will be greater at the poles than at the equator because the mass is closer to the centre of gravity of the Earth at the former. Clearly schoolboy physics apply to an homogeneous body, which the Earth patently is not, and sensitive instruments have been devised to measure very accurately the local variations of gravity due to small-scale density variations in order to provide information for the geologists and geophysicists.

The solid surface of the globe comprises rocks which are in a constant state of transition, either moving relative to one another as in the case of the tectonic plates, or being subjected to intense and immense stresses and strains not only by their relative movement but also by such factors as temperature differentials, pressure, volcanic action, etc. Throughout the changes caused by these stresses, the rocks retain what, in simplistic terms, may be referred to as a memory of their history. Of this memory, one of the more significant aspects is that of magnetism. Geologists know that throughout time - or at least the 4,600 million years of the Earth's existence - the relative positions of the north and south magnetic poles have changed many times. The rocks that form the Earth's crust have locked in the magnetism that was in being at the time that they were formed, and thus offer evidence of both the timing of their formation and of their relationship to adjoining geological structures. The differential magnetic signals locked into the rocks provide one of the ways which geophysicists use to measure the relative movements of the tectonic plates.

The two Precision Echo Sounder systems (PES) are used to measure the depth from the surface to the seabed, but in slightly different manners. Both emit short, sharp pulses of sound, one at a frequency of 10 kHz (10,000 cycles per second), the other at 3.5 kHz. The higher frequency sound suffers greater attenuation in its passage through the seafloor sediments, and thus its reflection is essentially only from the actual surface of the seabed - ie it provides a measure of the depth of the sea at the point of measurement. The lower frequency sound suffers less attenuation, and it retains sufficient energy to allow it to penetrate, and be reflected from, interfaces within the sediments that lie beneath the seabed itself. Dependent upon a number of variables which range from the state of the sea to the composition of the actual sediments themselves, the 3.5 kHz PES can provide an image of the top 30-80 metres of sediment which overlay the volcanic basement rocks beneath the seafloor.

Further imaging is obtained by use of seismic signals generated either by airguns or explosions. Both sources generate energy at a much lower frequency than the PES already mentioned, and their signals are thus able to penetrate beneath the seabed to a much greater depth - many tens of thousands of metres. With airguns, the frequency of the source can be controlled by using guns of different sizes - usually measured in terms of cubic inches - and sophisticated control systems have been developed to fire an array of guns in a particular sequence in order to maximise a preferred frequency signal.

The intensity with which each potential drilling site was surveyed may be judged from Plate 7 opposite, which shows the ship's track for two of the sites (designated CARB-1 and CARB-2). As a result of these surveys, three of the six originally selected sites were deemed suitable for drilling, and an alternative to another was selected after analysis of the cruise data. In total, geophysical data was collected for over 470 nm of ground in 82 hours. This part of the cruise was judged a success.

The commencement of the second phase of Dr Baxter's work (a survey of the Rodrigues Ridge) was, however, delayed by problems with the MCDSAS, which caused headaches for the technicians on board for just over two days - and, of course, two days on board means 48 hours non-stop work to get the system operational and fully effective. Once the system was fixed, it performed almost flawlessly, and over 700 nm of multi-channel data were collected together with some 1,250 nm of magnetic, etc., data as for the first part of the cruise. Using initial analyses of some of this data, the scientists made a number of core, dredge and trawl deployments, all techniques aimed at sampling the sediment overlaying the basement rocks.

The basic corer is simply a long barrel with a heavy weight on top. This is dropped over the side of the ship on a long, but slack, wire. The freefall of the corer drives the barrel into the surface of the seabed, and the coring tube is then closed at its lower end to retain the sample. When it is withdrawn by the wire it is filled with material which reflects the composition of the seabed at that point. The core is then recovered by the ship for subsequent analysis and archiving. Colour plate 4 shows a typical core sample opened up for analysis. Of course, the length of core recovered will depend on the extent to which the corer penetrated the surface, but a short core indicating a rocky surface is as significant as a long, sediment-filled, sample. The corer achieves its ultimate current development in the Ocean Drilling Program, where a sophisticated drill is driven several tens of metres into the seabed to recover cores which can reveal the longer-term geological history of the area.

Dredging and trawling are different forms of a technique whereby an open-mouthed container is dragged along the seabed to gather loose rock samples. Finally, a further technique uses a grab to collect samples from a specific area. Dredging proved to be exceptionally difficult on the rocky basalt surface of the Rodrigues Ridge and eventually one of the dredges was lost together with 3 km of cable. The scientists were reduced to manufacturing ad hoc dredges from empty acetylene cylinders.

Scientifically the cruise was highly successful. Despite the dredging difficulties the ridge was well sampled. Analytical data show the ridge to have some unusual geochemical features, resulting from interaction between the Reunion hotspot and the Central Indian Ridge.

Cruise CD23/87 was led by Dr Lindsay Parson, from IOSDL. The formal remit of his cruise was:

> ..to undertake a geophysical survey of the Indian Ocean Triple Junction (in the vicinity of 25°S 70°E) and parts of the surrounding seafloor. Specific objectives are to obtain GLORIA data on spreading axes covered by previous French SEABEAM surveys; to completely insonify the youngest ocean floor generated at the slowest of the three axes (the Southwest Indian Ridge); to study the morphology of fracture zones close to the Triple Junction; to compare the character of each of the intraplate

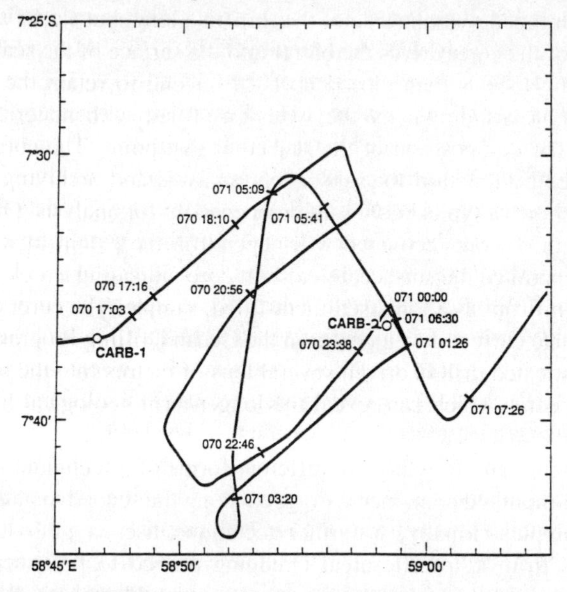

Plate 7: Map of CARB1 and CARB2 surveys ex Baxter cruise report

Triple Junction traces and to obtain air particulate samples throughout the cruise.

RRS *Charles Darwin* departed the Seychelles on 13 May 1987 and arrived at Mauritius on 11 June. Between these dates, after an initial period of establishing the serviceability of the equipment whilst within territorial waters, equipment was deployed to collect the various types of data necessary to meet the objectives of the cruise. The formal cruise report makes interesting reading, particularly if one reads between the lines, and no excuse is offered for quoting it below.

> Data logging commenced at 1425 14 May (134)[11] . Initial results from the 3.5KHz profiler were poor at a survey speed of 10 kts but the range of GLORIA was up to 30 km full swath, and clear insonification of slides and debris flows off the east flank of the Saya de Malha Bank was obtained. Fine weather and calm seas took us southwards in a series of dog-legs along the Central Indian Ridge (CIR), during which we insonified several charted and uncharted fracture zones and ridge offsets. The seismic reflection profiling system was first deployed 0800 19 May (139) to complete the southern section of the CIR at 8 kts, but recovery was necessary 0445 21 May (141) after a ripped trigger lead caused gun failure. It was decided to recover the streamer in the light of the small amount of sediment observed, and the survey continued at 10 kts. A series of north-south lines were set at approximately 16 nm spacing to cover the triple junction box, and these were completed by 0620 25 May (145). Between 0632 and 1747 24 May (144) we had to break off the track line to take emergency evasive action in avoiding collision with an uncommunicative tanker. At 1017 (144) we once again had to reduce speed to 4 kts to avoid collision with another vessel. (Parson 1987). As the Master records of these events 'Clearly even the remotest waters are not safe from cowboys.'

The cruise report continues in this vein for some time, with the odd buried sentence that enlivens the meticulous prose - eg 'Course at 063 degrees was held with weather suitable enough for the mid-cruise barbecue...' Note the definite article - 'the' mid-cruise barbecue, not 'a'

- as though it were a well-established tradition, which indeed it seems to have become. Colour plate 6 shows a mid-cruise barbecue and colour plate 5 the sort of view over the bow that every mariner dreams of; the adjoining pictures may represent a more common occurrence, and are referred to in a later chapter. But perhaps this supposed tradition is an essential part of maintaining momentum in what could otherwise become the boring routine of data collection. Certainly the cruise report notes that '..twenty six days of GLORIA survey were completed with only two breaks...Apart from these interruptions, data was logged continually and was unaffected by the few problems that did occur'. Some of the boredom was relieved by the intellectual effort of writing suitable contributions to *The Beagle Times*, and one limerick that resulted was

There once was a vehicle named GLORIA
Towed deep, mapping oceans' broad flooria.
The triple junction was best,
Looked so like a nest
That young Lindsay near died of euphoria

The cruise gathered data which formed the basis for a number of papers. The Indian Ocean Triple Junction is a region of significant geophysical interest because, as its name implies, three lithospheric plates meet there. The Indian/Australian, African and Antarctic plates intersect, with two pairs moving apart at about 5cm per year, and one pair at only about 1.3 cm per year. (The average movement of all tectonic plates is about 3cm per year, about the same rate as the growth of fingernails.)

On the basis of GLORIA data gathered on this cruise, Parson *et al* (1993) put forward the theory that the Central Indian Ridge was divided into segments by what they titled 'ridge-axis discontinuities'. These discontinuities were not regular, and in a curiously unscientific terminology, the paper suggests that they were 'controlled by variations in magmatic plumbing'. Mitchell and Parson (1993) also developed the GLORIA data to posit that the Indian Ocean triple junction evolved in two ways - one continuous producing classic scarps and herringbone intersections, the other discontinuous with the *en echelon* faulting referred to in the paper cited above. Both papers drew attention to the similarities and differences between the Indian Ocean and other triple junctions throughout the world's oceans, and both suggested that the theories would repay further study. Hence, Parson and his co-workers postulated further research using complementary techniques to attempt to answer some of the questions raised, thus re-inforcing the old adage that any good science raises more questions than it answers.

---

[6] A fire-suppressant foam, called Hi-Ex because its constituents expand rapidly when mixed in use.

[7] facies - The aspect of a rock unit in terms of the conditions under which it was formed.

[8] Phytoplankton - plant-like plankton, as opposed to zooplankton

[9] Victoria A Kaharl, *Water baby - The story of Alvin* (New York: Oxford University Press, 1990) 350

[10] Peter Gurney, *Braver men walk away* (London: Harper Collins 1993) 250.

[11] The time reference used on scientific cruises is normally Greenwich Mean Time and Julian Day. The former is obvious but the latter refers to the serial number in the year of the day in question - ie January 1st is Julian Day 1, February 1st is Day 32 etc. For international comparisons and later analysis this system avoids potential ambiguities generated when working in changing time zones, particularly on later cruises where operations were carried out near or across the International Dateline.

4
Physical Oceanography

Physical oceanography was the discipline which attracted considerable attention from the university community, and was perhaps the most intensively international topic. Two cruises were led by scientists from the University College of North Wales (UCNW), two by the University of Miami, one by Woods Hole Oceanographic Institution, and a sixth by a team from The Queen's University of Belfast. The obvious interest was generated by the effect of the monsoons in the area upon the physics - currents, water temperature, etc. - of the coastal waters of the Indian Ocean and the Arabian Sea, but the studies encroached upon what may be termed 'biology' because all involved some assessment of the impact of the physical parameters on biomass production.

For the cruises led by UCNW, the scientific objectives formally stated were:

> ..to determine the modification of the flow and any increase in vertical mixing induced by the presence of an island in a strong mean flow. Two particular objectives will be to identify any trapped or connecting eddy systems in the wake region behind the island and to assess the extent to which the island-induced effects contribute to biomass production. Research will concentrate on the waters in the vicinity of Aldabra.

These two cruises arose from genuine scientific curiosity. Biologists had long suspected a phenomenon which they had christened the 'Island Mass Effect'. Put briefly, this resulted in a richer biomass on the downstream side of an island. The research around Aldabra was aimed at finding whether this effect could be quantified in physical terms. The two cruises, separated by three months, allowed the deployment and subsequent recovery, of moored instrumentation to ascertain whether there were significant changes in currents over the period. Heywood *et al* (1990) conclude

> ... The results are consistent with the theory, in that at moderate speeds an eddy appeared trapped in the lee of the island, while at lower speeds no eddy activity was observed. The current speeds were too low to expect a wake of eddies to be spun off behind the island.
>
> On those occasions where an eddy did appear trapped behind Aldabra, a significant effect was observed on the hydrography and biology. The isopycnals[12] rose in the lee, giving an area of colder sea-surface temperature, associated with enhanced chlorophyll concentration due to greater phytoplankton productivity. Cooler water has been brought from below the surface, bringing with it nutrients which increase productivity. Therefore if an oceanic island is in a region of strong, steady current, the island mass effect is an important contribution to the productivity and potential for fisheries in the area.

In other words, the 'Island Mass Effect' was borne out by measurement, and an immediate potential commercial application noted. Of course, it is likely that fishermen who have lived on small islands for years have exploited this effect without the benefit of scientific measurements, but with the reward of consistent catches in particular waters! But for the future, the conclusions of these two cruises may be used to guide fisheries development around uninhabited islands, when and if this proves necessary.

One of the primary tools used on these cruises was the Acoustic Doppler Current Profiler (ADCP). This instrument measures differential currents in the ocean so that flow rates at different depths can be assessed. It does this by generating an acoustic signal from a transducer mounted (in the case of RRS *Charles Darwin*) in a housing underneath the hull, and the phase-shift of the returning 'echoes' gives a measure of the speed and direction of subsurface currents relative to the ships motion. The theory (and the name) originate with Doppler, who formalised the perceived tone differential observed when one sound source is moving relative to the observer - the most common observation is that of a train passing an individual.

At the time of this cruise, the system was still fairly novel to NERC-supported scientists, although Professor John Simpson, the Principal Scientist for CD22/87, had been one of the leading proponents for its purchase. Nevertheless, Karen Heywood records in her diary

> ..I spent the morning with Ken trying to fathom out the Acoustic Doppler Profiler. It turned out that John Simpson and I were right, in that when taking a mean as reference level, the ADCP took the mean velocity of the profile away from everything (not the velocity at the mean depth, as Ken thought) But we can't work out what it does when it references to a zero velocity level. One would expect the velocities to return to zero at the prescribed reference level, but they aren't unless the ship is stationary. We've sent telexes off to the manufacturers in San Diego to see if they can explain what is happening.

That Karen was able to work out what the ADCP did is shown by the fact that on CD24/87, she became the 'expert' on the ADCP, and noted that she was able, albeit hesitantly at first, to give a presentation on it to the new scientists aboard.

However, on these two cruises an alternative use was found for the information returned by the ADCP. Instead of using only the principal signal, the phase shift of the backscatter was measured, and compared with the abundance of zooplankton in the water column. Heywood *et al* (1991) report:

> Significant correlation is found between the range-corrected backscatter summed from the surface to a depth of 200m and the

biomass collected by a net haul over the same vertical range at the same station....

....the application of the ADCP to combined studies of flow fields and biological productivity offers promise of continuous along-track estimates of zooplankton biomass when supplemented by spot calibrations.

Thus, the scientists opened up a relatively new area for study by applying lateral thinking to data collected by an established instrument. In such manner does science expand, and as already noted, raise more questions requiring further study. The cruises also generated a new verb.. to 'Dopple', defined as activation of the Acoustic Doppler Current Profiler.

But the scientist is always seeking to expand his horizons at a time when his ambitions may be (in his view) unduly constrained by funding limitations. These two cruises illustrate graphically two aspects of this comment.

The first is that Professor John Simpson and Dr E Des Barton (based upon work done by Simpson in shelf seas) initially put in an application to NERC to study the effects of islands on current flow in the Atlantic, but were advised by the review committee that the submission did not merit support. It was later suggested to them that the presence of RRS *Charles Darwin* in the Indian Ocean offered an ideal opportunity to study the effect of Aldabra Island on current flow. Subsequently, research by the two UK scientists in collaboration with European colleagues on the flow effects around Atlantic islands has revealed a more marked effect than that from Aldabra - I suppose the reviewers can't be right every time!

The second (or perhaps it should be the third, given the comments about the application of the ADCP data) was that, to quote one of the principal scientists, they 'stole' three or four days from their approved cruise time to monitor currents through the Amirante Trench. This resulted in a paper (Barton & Hill 1990), which was not an expected end product of the cruise, but which in fact had a significant effect upon the direction of the (then) nascent World Ocean Circulation Experiment (WOCE) programme.

The UCNW scientists were able to combine their professional visit to Aldabra with a brief mid-cruise period of relaxation. Aldabra Island is, strictly, an atoll - an area of ocean almost totally enclosed by land, about 5 miles wide by 10 miles long. The enclosed water is tidal, and the scientists from the ship wished to measure the tidal currents through the channel, so the ship anchored off Aldabra and the scientists prepared to go ashore. They did so with some trepidation, because '..we heard all about our day on the island of Aldabra - apparently there are nasty vicious land-crabs, Black Widow spiders, and, we reckon, probably sharks. Also there are lots of mosquitoes, and the coral is very sharp and will cut your shoes up'. In the event, it was only the last of these expectations that was justified: the island was volcanic in origin, and therefore the surface was marked by irregularities and fissures which made walking difficult - see Colour plate 8.

Once on the island, they were welcomed by the local inhabitants (a scientific team of about 8 people, and a small indigenous group), and made to feel welcome. So much so that after a relaxing swim (with snorkel and flippers) in the channel, where the moderate current enabled the swimmers to drift effortlessly on the surface and observe fishes from the brightly coloured darting species near the surface, through small 'ankle-nipping' sharks in the mid-depths, to the lazy 'flying' of the rays just above the bottom, scientists, crew and locals got together to enjoy a lunchtime barbecue comprising principally fish. There was a humorous sideline to this event. Karen Heywood's diary notes that

> ..this was a source of some embarrassment, since there'd been a fiasco with the sandwiches provided by the Darwin. The cook had got rather drunk the night before, and was cheesed off because he couldn't go to the island, so he'd thrown the bread overboard. Then Sam (Mayl), the Captain, insisted he make more bread, so he was up all night doing that. When we arrived on the island and found lunch was to be provided, we didn't dare not have the sandwiches - so we gave them to the Aldabrans to put in their freezer!

At this stage, it is pertinent to mention an event which occurred on several cruises, namely the catching of a shark. It appears that almost anyone who had once used a fishing line had ambitions to catch a shark, and many were successful. However, the means and eventual end of the story were so different on different cruises that they sit well as a whole. The UCNW team were obviously fishermen in the classic mould - their shark took the bait, but as they were hauling it aboard, it slipped the hook and escaped. We are not told how big it was, but almost certainly with the telling it will have been bigger than any other. Nevertheless, not everyone was happy at the idea of catching a shark, and one scientist records 'I was greatly relieved to hear that it had slipped off the hook. I was upset at the glee with which the sailors captured this beautiful wild animal. They didn't want to eat it, just to kill it. Horrible'. The counter-view was put by another scientist who wrote 'Anyone who knows of people killed or maimed by sharks can understand, if not entirely share, the blood lust that they bring out in sailors'.

On another cruise, the scientists observed a sizeable shark swimming off the port quarter and snapping up anything that appeared. This apparently included an empty sardine tin, and so a plot was hatched to catch this shark. Another part-empty food tin was dangled in the water, adjacent to a rope with a noose at its end. By some chance, which only happens in fishing stories, the shark was persuaded to attack the tin by putting its head through the noose. Once this happened, the noose was tightened, slid down the body of the fish, and was then snagged by its tail. The rope was led over the ship's A-frame, and the shark - estimated at some seven feet in length - hauled in over the stern, where it was left hanging for 3-4 hours. Someone suggested that shark steaks would be rather nice for supper, and the cook, bearing a sharp carving knife, approached the now apparently moribund shark. It appeared that the shark was not as moribund as had been supposed, for when approached closely, it moved and snapped its jaws. The cook was reluctant to try to prepare steaks from such a situation, and so the shark was left hanging for another couple of hours. After this period - some 6 hours - everyone assumed that the shark must be dead, because it is received wisdom that sharks 'have to keep moving to stay alive', and presumably this was even more true when it was suspended in air. However, nobody had told the shark this simple fact, and an approach by the knife-bearing cook again resulted in a twist of the body and a snap of jaws. At this, there was agreement between all involved that the shark had suffered enough, and so the rope was cut and it fell back into the water. With a shake of its head, and a flick of its tail, it swam away from the ship, apparently none the worse for its adventure. And so, if at some future date, a fisherman or research biologist catches a shark wearing a noose around its tail, he will now know how the rope got there.

Finally, of course, it is necessary to relate that on at least one occasion, the shark did not get away, and shark steaks appeared on the evening menu for a couple of days. These were supplemented by flying fish served in a variety of ways, because in the tropical latitudes these fish are highly mobile, and were often found upon the after deck of RRS *Charles Darwin*.

On both CD22/87 and CD24/87, there were periods of intense and prolonged activity - such as when all the data had to be catalogued and logged on magnetic tape immediately prior to the end of the cruises - interspersed with quiet spells, such as when parts of the scientific team were off-watch. To fill these periods of idleness, not all of which could be spent in sleep, in the bar, or watching videos, various activities were devised, of which one of the more popular was skipping. However, one of the scientists notes that this was not so much to while away the idle hours as 'some form of keeping in shape against the onslaught of rich, tempting and overgenerous helpings of ship food, and was effected strictly between 1600 and 1700, as I recall'.

Deployment of recording instruments in the ocean requires, by definition, huge quantities of rope of varying sizes, short lengths of which are often to be found aboard and make great skipping ropes. Not only was it found that skipping offered an ideal form of physical activity in the limited space offered on board ship, but it soon became a form of competition, with informal marks being awarded for numbers of uninterrupted skips, aggregate skips, and technical competence with manoeuvres like cross-hands, bumps, and backwards skipping.

And then time which might or might not have been idle could be filled by 'Board of Trade Sports', the sea-goer's affectionate term for safety training. How this was effected was a matter for the Master, and this is reflected in a comment shown to the writer. 'At 0830 we had an hour's talk from the Captain about safety. He isn't the same Captain as last time, and he's rather a stickler for the rules. Treats everyone like schoolboys and thinks he's the headmaster'.

But then, when 39 people of widely different backgrounds are constrained to live together for 30 or so days and nights, some impressions and preconceptions are bound to form. Karen Heywood records in her diary:

> ..They played a nasty trick on me and Ray this morning. We have adjacent cabins, and each person's cabin has their name on. This morning someone swapped over names, so that I went into Ray's cabin by mistake. Poor Ray, he was fast asleep since he's on the 12-4 watch, and I woke him up. I did feel rotten. Of course, then they started spreading rumours that I'd molested Ray in his cabin, and I got a lot of teasing...Then later in the cruise, ..After I came off watch (about midnight, since Karen had the 8-12 watch) last night, I went into the bar to unwind (just drank water, though), and talked to the others - Phil, Des, Graham and Scrope. Mostly they talked about beer and pubs - it was a bit boring and I didn't join in, but it's interesting to see what men talk about as a group. Must be very dull. Went to bed about 1 am.
>
> From such occasions are lasting impressions generated.

This cruise ended in Mombasa, and Des Barton recalls two events. The first was on the approach to Mombasa, when, at about 5 am he was awoken by the officer on duty to ask permission to switch off the ADCP as the ship was about to leave international waters and had no permission to work in Kenyan waters. Des commented that 'the danger of being challenged off the East African coast for illegally operating a Doppler Profiler seemed somewhat remote. Although normally I am a patient and tolerant person, his early call was not appreciated the morning after the traditional RPC'[13]

When the ship docked in Mombasa, Des notes 'everyone was desperately keen to get off the ship after several weeks at sea'.

> To the scientists' horror, the medical authorities of the port, as represented by a rather fearsome nurse, stated that we must all be injected against cholera before being allowed ashore because there was some outbreak in the region. Since everyone had had every conceivable variety of painful arm- and backside-numbing, lump and fever-inducing injection, sugar cube and inoculation prior to leaving the UK as well as swallowing varieties of anti-malarial pills of dubious effectivity throughout the cruise, no-one was particularly keen on being re-injected against cholera. Some

were close to sedition because of the rising awareness of AIDS transmission by poor needle hygiene in East Africa. Though we had all been given protection against cholera, the Llandudno Medical Centre had assured us that no documentation was necessary anywhere in the world as it was comparatively rare.

The nurse was not impressed. No paper - no going ashore. Nobody knew whether being inoculated twice at such a close interval might have any adverse effect. Advice sought from the Captain brought a typically helpful response that it was nothing to do with him and he had no opinion (*a realistic stance, since he had not been given in-depth medical training. FPV*). As Chief Scientist I had a word with the nurse, who remained adamant. I then had a quiet word with the ship's agent to see if there was any other way of resolving the difficulty. Like ship's agents the world over, he had a helpful suggestion. So, armed with a bottle of the best Cognac and several cartons of cigarettes from the ship's bond, purchased at great personal expense, I was escorted to an interview with the Chief Medical Officer of the port. There I flourished a letter on headed paper covered with lots of ship's stamps stating the importance of our scientific mission, the great appreciation we had of the co-operation offered by the nation to our work, and that due to our administrative oversight we had no cholera certificates, and as a token of friendship could we present some small gifts. No problem - everyone off the ship like a shot and into R & R pending the flight home. And no-one caught cholera.

The American-led cruises covered the two monsoon seasons in that part of world - the North-easterly around the turn of the year, and the South-Westerly during the northern hemisphere's summer. Both cruises were led by Professor Don Olson, whose past included a spell as deckhand on a Mississippi paddle steamer. Currently based at the University of Miami's Rosenthiel School of Marine and Atmospheric Sciences (RSMAS) on the causeway between Miami city and the Florida Keys, he had as co-Principal Scientist Dr Bruce Warren. Bruce was born and raised in Boston, Mass., where his family still live, and he has only moved some 90 miles from there to Woods Hole Oceanographic Institution (WHOI) where he has been for 30 years or so. The contrast in appearance between the two men is striking - Olson, bearded, extrovert, whose office displays the 'volcano theory' of filing on every available space, Warren almost patrician, pipe-smoking with a clearish desk and a chair available for the visitor - but they share a lively outlook in scientific terms, reflected in the fact that they had obtained funding for this cruise on, what was to them, a 'foreign' ship because it offered the opportunity to measure the effects of the monsoons on currents in a little researched area of the world's oceans.

The official description of the science to be undertaken reads:

> ...to undertake physical studies of the waters of the western Indian Ocean - to include mixing layers, water mass analysis - to assist in the understanding of the relationship between the changes in the surface velocity field and the large signal in thermocline depth seen in the basin over the monsoon cycle.

Warren and Johnson (1992) report :-

> Mapping of the waters present in the deep-water layer above the bottom water during successive monsoons was undertaken because three earlier surveys had shown differing horizontal distributions there, and had raised the possibility of a monsoonal reversal of the circulation, paralleling that in near-surface water. However, the patterns of variation in the salinity and oxygen fields were essentially the same on the two *Darwin* cruises, suggesting a broad southwestward flow in the deep water layer of the Somali Basin during both monsoons. We propose that this pattern was not forced by the monsoons but reflected mainly the mean circulation of the deep water as driven by the upwelling of the bottom water from below, that that pattern might be enhanced during the northeast monsoon if the wind-forcing penetrates into the deep water, and that a moderately strong southwest monsoon (in contrast to the extraordinarily weak monsoon of 1987) would be required to disrupt the distribution of water properties.

Thus, Warren and Johnson's findings upset theories existing at the time, and led them to advance the view that the deep water currents of the Arabian Sea were not affected by the monsoons. In another paper arising from these cruises, Johnson, Warren and Olson (1991) develop the theory that the bottom water in the Arabian Sea is supplied primarily through the Owen Fracture Zone in the Carlsberg Ridge. To arrive at these conclusions, the researchers used measurements of salinity and temperature at upwards of 100 'stations' (a station being a predetermined point in the ocean where the ship is halted while a recording instrument is lowered to a specific depth, and samples of the water collected in sampling bottles at various depths as the wire is wound in) on each cruise, and these data were then correlated and compared with existing information and theories to produce the papers which would take the study forward. As an indication of the detail required for a study like this, seawater temperature was recorded accurately to +0.002°C (or an accuracy of better than 1/500th of a degree), and salinity to better than 1 part in 100,000.

The cruise began with a worrying period when airfreight from America had not arrived in Oman. Since one component in that airfreight was the IBM Personal Computer (PC) used to analyse and display the deep water currents which were the subject of the cruise, there was much nail-biting in the day or so before scheduled departure. Some of this anxiety was forgotten when Olson, Warren, and a couple of the technical support staff were invited to a reception at the British Embassy, adjoining the palace of the Sultan of Oman. Olson notes '..British Embassy dates back to the Colonial period.....An interesting note is that there is no word from the US Embassy. Either the State Department has still not passed the information on the cruises or they just don't care.' Ken Robertson from RVS, Barry, also attended this reception, and he supports the American view:'...this magnificent old building in a beautifully scenic position overlooking the sea. A finger buffet and drinks had been laid on and the whole occasion reminded me of some Somerset Maugham or Noel Cowerd (*sic*) novel setting. Discreet waiters appeared each time one's glass was empty and the two hours or so was a most pleasant experience so far from home.' In the event, after much telephoning between Oman and Barry, the American air shipment was located at London's Heathrow airport, whence by sleight of hand it was transferred to another airline and shipped to Oman, where the four boxes arrived some 2 days later than planned.

The ship sailed from Mina Qaboos on 20 December, and although Christmas loomed work got underway immediately. As is normal on such cruises a watch schedule was implemented, but Olson was sufficiently concerned to note in his diary ' ...need to get the NERC technicians on a schedule that doesn't kill them'

However, the diary also notes

> ..Various groups, with urging from the chief cook... are decorating the ship for Christmas. We have lights, ornaments and balloons, supplied by the cook, in the main laboratory.... Roger (Chamberlain), the Second Officer, has come up with a game for Christmas. The idea is a murder mystery. Everybody on board picks a card from a deck and is given two chestnuts. Then the recipient of the Queen of Spades is the murderer. He or she can only kill someone if they are alone. The victim has to give up their nuts. The living have to try to identify the murderer.

The first 'murder' was at 0002 on Christmas Eve, and another (of Russell Griffiths, a NERC Technician) took place in the shower. Since the 'murderer' turned out to be Lucinda Hubbard, from the University of Belfast, one has to ask what she was doing alone in the shower with Russell in the first place. However, the essence of Christmas highlighted George Bernard Shaw's comment that the British and Americans are 'two nations divided by a common language'. For the evening of Christmas Day the ship's stewards had provided a buffet and everyone was invited to come in fancy dress. The UK participants duly arrived in a variety of costumes made up from whatever they could find on board; the American scientists arrived in jackets, collars and ties. However, any temporary embarrassment was more than overcome by the efforts of the ship's stewards, who appeared in Father Christmas costume, delivering presents to all aboard from their sacks, an event remembered fondly by the American scientists.

Nevertheless, science was suspended for only a part of Christmas Day itself, and the diary is littered with references to CTD casts and recoveries taken at midnight, 0200, 0400 and similar hours unthinkable to the landbound scientist. Some of the excitement of the research shines through this record of events:

> ...Pattern is extremely interesting with high salinity Arabian Gulf water extending down to 11-12°N, then a transition to fresher mid-column, a blob of Bay of Bengal water (?) from the Indian coastal current and then waters of Southern Hemisphere origin in a broad (200-600m) salinity minimum. Below that is a broad salt maximum from the Red Sea.

But lest it be thought that scientists' only interest is hard fact, Olson also records a group standing on the after deck, watching the sun go down and waiting for the 'green flash', a phenomenon peculiar to equatorial latitudes. Warren and one or two others claim to have seen it. Olson professes some scepticism, but then goes on to liken the crescent moon with a bright companion star to the Turkish flag. Later, he writes of the moon making ' ... highway of light in the sea. I try to capture her glory on film'. Tony Cumming, an RVS Technician who was not on this cruise could, perhaps, have settled the green flash argument once and for all. He claims that it is a well-established phenomenon in tropical and sub-tropical latitudes, and can always be seen when the sea is calm and there is a stable high pressure system over the area, so much so that (he claims) it is possible to judge the second or so at which it will appear precisely enough to focus a pair of low-power binoculars on the final crescent of the disappearing sun in order to get a better view of the 'green flash'. Karen Heywood, on an earlier cruise, had noted ' ..After dinner, the mooring was in its final stages. We all watched the sun going down - as it goes below the horizon, you are meant to see a small green flash of light. I couldn't see it, neither could I see it on the video we took, but several people swore they did'. Perhaps this too offers an opportunity to extend marine research into a new topic area.

This cruise was the first of many to 'Cross the Line', and the event was celebrated with due ceremony. Let Don Olson tell it:

> Neptune arrived at 1400, with a vast array of minions - a Norseman, policemen, surgeons, Keith (the Chief Officer) as Master of Ceremonies, the bears, the barbers, and of course Aphrodite... They assembled on the foredeck and hailed the Captain...Neptune's herald addressed the bridge, demanding to know who commanded the vessel. In due course, Neptune's people were welcomed aboard (see Colour plate 9).

They then proceeded to the stern where the throne overlooked the pool for the bears; the barber's chair and the surgeon's table had been laid out in preparation.....Basically, the proceedings commence with the policeman bringing the 'polly wogs' to the front, where their sentence (*the word* sentence *is the US equivalent of* charge *in UK English*) is read. They are then pronounced guilty, and 'examined' by the surgeon. This involved an operation in one or two cases, but in the majority there is only the simple administration of medicine (one or two doses). The medicine is a mixture of salad dressings, lemon and orange juice, and red food colouring. The victim is then passed on to the barbers for their punishment. This consists of a mock total shave. Lots of 'shaving cream' is administered, followed by a going over with a huge 'razor and comb'. In a couple of cases they also cut up a piece of steak with scissors and dropped the remains on to the person's neck to simulate actual trimming. This is all followed by the administration of 'aftershave' or 'perfume', depending on sex. This is given in flagons, and consists of galley slop, fermented for a few days in a trash can. It is light yellowish-brown and extremely foul smelling. This goes alternately down the front, in the pants and over the top of the head. Finally, the victim becomes a 'shell back' with immersion in the pool with the bears.

It is interesting to note that Charles Darwin records a very similar event in his diary, where he begins 'We have crossed the Equator, & I have undergone the disagreeable operation of being shaved'. He subsequently records a very similar process to that noted by Olson, commenting 'that the whole ship was a shower bath, & water was flying about in every direction: of course not one person, even the Captain, got clear of being wet through'.

Such, then, is the serious nature of 'crossing the line', relatively unchanged over the years, and the ceremony, which was doubtless performed many times during the global voyage of RRS *Charles Darwin* will not be spelt out in full again.

One other aspect of general application to all of the cruises was particularly noted and enjoyed by Olson. It is a statutory requirement on NERC ships that regular safety drills are carried out. As a minimum this involves familiarising scientists with their lifeboat stations, the use of lifejackets, and operation of the lifeboats. Beyond that, further training is at the discretion of the Master - the 'Board of Trade Sports' mentioned earlier. On CD19/86, this included fire drills, where Olson records his boyish delight at being allowed to 'put out a fire' in the chemistry container, getting thoroughly wet in the process. On CD21/87, one of the RVS support staff used his expertise as a First Aid trainer to instruct some of the scientists and crew into the 'Buddy' system of emergency support.

The final physics cruise in this part of the world was led by Dr Graham Savidge, of The Queen's University of Belfast. Unusually, this cruise was of only 10 days duration, but it had been commissioned by the Omani Council for Conservation of the Environment and Water Resources (CCEWR) to:

> ...investigate the importance and spatial distribution of upwelling off the southern coast of Oman, during the south-east (*sic*) monsoon regime, and to assess the resulting perturbations in the physical, nutrient and chlorophyll fields. It is intended also to derive basic current flows, both from consideration of the physical data and directly from current drogues.

Savidge *et al* (1988) and Elliott and Savidge (1990) reported evidence of upwelling associated with headlands of Oman and an associated narrow filament of cool water streaming seawards from the centre of the upwelling. Their observations supported remote sensing imagery which suggested regions of intense upwelling close to the Omani coast - the upwelling water showed a temperature at least 5°C lower than that of the ambient water. Since such upwelling brings to the surface waters rich in nutrients, particularly nitrogen on which phytoplankton and thence the higher order food chain can flourish, the findings have considerable significance for the fishing industry of Oman. In the *Times of Oman,* August 20 1987, Dr Savidge is reported as saying

> Results from the surveys will enable further assessment of the ecological resources in Omani waters and provide a good background for further studies. Savidge *et al* (1988) report that 50% of the world's harvest of fish originates in these upwelling regions. The economic importance of the upwelling regions is far

out of proportion in comparison to their area....coastal upwelling zones cover only approximately 0.1% of the surface of the oceans of the world. The majority of these regions are located on the eastern margins of the major oceans and include the Californian and Peruvian systems together with their Atlantic counterparts the NW African and Benguelan systems. The one significant exception to this generalisation is the north-west Indian Ocean, where marked upwelling occurs on a seasonal basis in response to the monsoonal wind regime.

The basis of the Savidge cruise is significantly different from almost all of the other cruises in the global circumnavigation of RRS *Charles Darwin*, in that whilst most were prompted (and funded) on the maxim of satisfying scientific curiosity or of adding to scientific knowledge, CD26/87 was a cruise with a commercial objective. Savidge *et al* (1988) write..

The upwelling area off Southern Oman is one of the few upwelling regions of the world where, although the potential of the fisheries for commercial exploitation has been recognised for several years, this potential has not been recognised to the full. In any rational proposals that are forwarded with the aim of establishing a plan for the development of an exploitable resource it is essential to have an understanding of the factors controlling the resource stock. For an exploitable fish stock dependent on well-defined basic physical and biological processes for its existence, it is essential to have detailed information on the fundamental controlling oceanographic processes.

But lest it be thought that the cruise was driven by solely commercial motives, Savidge continues

A sound knowledge of the general oceanography of an area is also of essential value to other disciplines. Upwelling regimes are highly dynamic and a knowledge of water movement patterns which can be deduced both from direct measurements and from basic hydrographic distributions will be of considerable benefit. For example, a knowledge of current structure will be essential for any model capable of predicting the fate of a pollutant or the transport of sediments in the marine environment. Field measurements of currents when considered in the oceanographic framework are an important complement to theoretical predictive models of contaminant dispersal or sediment transport. An understanding of the major oceanographic features of a region also provides a basic framework which allows the ready establishment of intelligent sampling strategies for further more detailed studies at a later stage.

In other words, whereas those cruises which had as their primary objective the satisfying of scientific curiosity may, in the longer term, yield some commercial benefit (see the Kenyon geophysical cruise reported in Chapter 5), a cruise undertaken principally for commercial purposes will also yield data which will add to the sum of scientific knowledge about not only that specific area but also oceanography generally.

The final cruise to be reported in this chapter, although not the last chronologically in the area, is concerned with biogeochemistry. Dr (now Professor) Mantoura of the (then) Institute for Marine Environmental Research (IMER), now the Plymouth Marine Laboratory (PML), led the first of these in September 1986. The area was chosen because it covered a wide range of biogeochemical features, geographically from the marine 'desert' of the mid-ocean through to the nutrient rich water of the shelf off Oman due to the deep water upwelling referred to above, and climatically covering the impact of the monsoon. The richness of the scientific opportunities is shown by the Coastal Zone Color Scanner (CZCS) shown in Colour plate 10.

One of the scientists aboard the PML cruise, with almost 20 years of marine research behind him at the time, has commented to the writer that in terms of scientific discoveries this particular cruise was the most exciting he had ever experienced. Some evidence of this view is conveyed by the special edition of Deep Sea Research (Burkill *et al* (editors) 1993) which was devoted to the results obtained on RRS *Charles Darwin* cruises CD 16/86 (September 1986), CD19/86 (December 1986), and CD25/87 (August 1987).

Fauzi Mantoura has made this author's job easier in that the opening of the *Cruise Report* (Mantoura 1991) is headed 'What were the main discoveries?' which question he then answers in three paragraphs reproduced below.

> The surface concentrations of NO_3 as measured by sensitive chemiluminescence techniques were found to be severely depleted to nanomolar levels in the sub-tropical gyre increasing by over a thousand times within the upwelling waters off Oman. These surface NO_3 levels exerted a major control on 'New' Production as measured by $^{15}NO_3$ assimilation techniques carried out using in situ incubation systems. Phytoplankton production in oligotrophic[14] waters was low (<0.3 gC $m^{-2}d^{-1}$) and dominated by <0.8 μm picoplanktonic cyanobacteria whose biomass contributed significantly to the particulated organic carbon (POC). Bloom production conditions of >2.6 gC $m^{-2}d^{-1}$ occurred in upwelling waters which were dominated by large (>5μm) diatoms. The vertical flux of sedimenting POC to the deep sea correlated with upwelling production, and this was ultimately responsible for maintaining the suboxic conditions in the intermediate waters of the north west Indian Ocean (NWIO). Bacterial abundance and production estimated using uptake of tritiated thymidine were high relative to primary production, and this required a significant flux of dissolved organic carbon from sinking particle flux to support this bacterial production.
>
> Throughout the NWIO we found extensive evidence for denitrification of NO_3 to yield volatile N_2O, a potent greenhouse gas, and a NO_3 deficit within the oxygen-depleted waters. We calculated that the N_2O ventilation fluxes from the NWIO could account for up to 10% of the global ocean denitrification to the atmosphere. Likewise, supersaturated levels of methane, another greenhouse gas, were found to derive from sedimenting phytogenic organic just below the chlorophyll maximum, and its ventilation in the upwelling regions could contribute a major proportion of the global marine flux of methane to the atmosphere.
>
> The concentrations of rare earth elements (REEs) and transition elements (Fe, Mn, Cu, Cd, Zn, Ni)[15] were profiled reliably for the first time in the Indian Ocean to reveal levels which were

intermediate between the Atlantic and Pacific Oceans, consistent with general ocean circulation and deep regeneration processes. Metals (Ce, Mn, Fe)[16] scavenged on to settling particles from the oxygenated surface zone were observed to be regenerated in the suboxic zones of the NWIO. Other metals (Cd, Zn, Ni and Cu) exhibited nutrient-like behaviour involving biological removal in surface productive waters and oxidative regeneration at depth.

Put more simply, these results indicated that this part of the Indian Ocean 'exhausts' to the atmosphere up to ten times more nitrous oxide than had been predicted. This would suggest that the oceans are as significant in the production of the (so-called) greenhouse gases as man-made sources. As one of the scientists on the cruise put it, this part of the Indian Ocean acted 'like one of the world's chimneys'. Conversely, the take-up of inorganic carbon (in the form of carbon dioxide) by picoplankton in the area may have contributed to the removal of this, another greenhouse gas.

These studies, and the results obtained, are thought to have had a significant influence on the subsequent structure and techniques of the

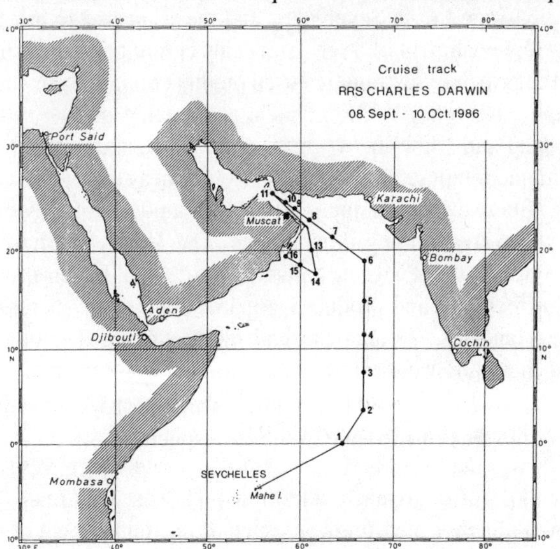

Plate 8: Map of Cruise 16, taken from Cruise Report

UK's 'Biogeochemical Ocean Flux Study' (BOFS) and possibly on the international Joint Geochemical Ocean Flux Study (JGOFS). Certainly it is a fact that the scientific team included (in addition to the UK scientists) representatives from the Netherlands, the USA, and Germany, whose Dr Hugh Ducklow and Professor Bernt Zeintzschel subsequently became leading lights in the guiding organisation for the US and German JGOFS exercises respectively.

This cruise yielded more than its fair share of anecdotes, some concerned with the science and others of a personal nature. The first of these can be traced back to the early days of preparation for the cruise, where equipment and chemicals were loaded into a container for shipment to the Seychelles to await the ship's arrival. Chemicals had to be packed to conform to International Maritime Dangerous Goods Regulations, and when the container was almost full one of the team approached Malcolm Woodward, who was in charge of logistics, and said that he had just remembered that he had to take a litre of sulphuric acid. Malcolm cannot remember the exact wording of his refusal of this request, but he was subsequently told by a third party that he was 'remarkably restrained'! An unfortunate conclusion to this story is that it was not recognised by some at the time that some of the chemicals would be adversely affected by the heat during passage to the Seychelles, and the team had to ask the local hospital for replenishment of a few chemicals before beginning the cruise.

One of the principal instruments used on this cruise was an Undulating Oceanographic Recorder (UOR), which, as its name implies, was towed behind the ship and 'flew' between the sea surface and a preset depth. It recorded depth, temperature, conductivity (a measure of salinity), and chlorophyll fluorescence (a measure of the amount of phytoplankton present in the water) among other things. The UOR was the particular 'baby' of one of the PML scientists aboard, and he ensured that it functioned effectively and produced good results for the majority of the cruise. Unfortunately, towards the end of the cruise, the towing cable parted, and the UOR was lost. *The Beagle Times* commented ' ...that X will be taking his future holidays on an Arab dhow in the Gulf of Oman, so that he can be near his beloved UOR!'

The UOR was deployed between 'stations', of which 16 were occupied throughout the cruise. At each station, a CTD cast was made, often in conjunction with other measurements such as the deployment of sediment traps. We referred to CTD stations in an earlier chapter, but sediment traps have not previously been mentioned. They are open conical devices which are submerged to collect the fallout of sedimenting particles. Within this conical housing a series of cuplike devices is driven into position beneath the opening for predetermined periods, thus giving a direct measurement of sedimentation rates to the intermediate and sedimenting depths.

These 16 stations were spread approximately evenly between the deep ocean and the shelf (see map in Black and white plate 8). Station 11 was the northern-most, sited almost in the Strait of Hormuz. At this time, the Iran-Iraq conflict was very much in the public eye, and *The Beagle Times* carried reports of ships attacked in the area. Station 11 was seen as essential by the Principal Scientist, as it was naturally closest to the shore and on the shallowest part of the shelf. As the cruise proceeded, 'Station 11' came to be used as a threat or a joke. The Master would give both crew and scientists advice on what to do, or not do, if boarded - 'Do not speak to the Iranians'; 'Do not laugh'; and various other worthy exhortations. In the event, Station 11 was occupied, but the scientists say that 'occupied' is too strong a word. What happened was that RRS *Charles Darwin* sailed into position, the CTD was deployed while the ship was stopping, and then when the CTD was being raised, the Master 'put the pedal to the metal' and sailed rapidly back to safer waters. The cruise report suggests that Station 11 was actually occupied for some 6 hours, but that figure has to be compared with occupancies of between 12 and 48 hours at other stations. The difference between actuality and recollection may be due to time and space being relative since Don Olson's colleagues prepared a banner proclaiming a 'Mine and Cheese Party', at the end of CD19/86, although that cruise went nowhere near the Strait of Hormuz! The banner hangs with pride over Don's office door in Miami.

RRS *Charles Darwin* suffered no ill effects from her expedition into the mouth of the Strait of Hormuz, and was able to continue her voyage and enjoy other adventures that awaited her.

---

[12] Isopycnals - lines of equal density

[13] RPC - Request the Pleasure of your Company, a euphemism for a party on the last night of a cruise.

[14] Oligotrophic - a body of water deficient in plant nutrients

[15] Respectively Iron, Manganese, Copper, Cadmium, Zinc and Nickel

[16] Ce-Cerium

5
Eastern Indian Ocean

The story of the next two geophysical cruises really begins in Mombasa some few days before the ship arrived. The science one - (CD20/87) a geophysical study of the Indus Fan - was to be effected with the aid of GLORIA, and the other (CD27/87) was to be a sampling and sedimentological study of the Fan based upon the GLORIA results.

As may be seen, the GLORIA package in its mount is quite sizeable - approximately 7 metres long by 3 metres high. GLORIA had been delivered to Mombasa, but unfortunately its most direct route to the quay alongside which RRS *Charles Darwin* was berthed was through an archway whose internal dimensions were less than those quoted above. It was, therefore, necessary to move the system into the shipyard via another gate. This happening is perhaps most succinctly recorded by the Telex from the ship to RVS.

From Charles Darwin sometime this morning.

A fraught afternoon eventuated in GLORIA plus two IOS containers in convoy to Yard. GLORIA got stuck under main entrance gate arch - exciting time by all. Finally brought convoy round to alternative access to Yard with police riding shotgun and totally ignoring detritus of telephone wires, street lights, tree branches and roof tiles that were demolished on route. Hotel diners interrupted by suitable noises off.

GLORIA etc. now on board - Vsl sitting in entrance to drydock whilst we finalise container sitting.

If Fisher[17] can put out lights in Wormley we do it better in Mombasa.

History does not record the views of the populace on this journey, which is perhaps just as well!

Even when GLORIA was brought alongside the ship, the drama did not end. The ship had undergone her refit in a yard outside the service quays and docks of the port area, and was tied up alongside a quay with limited craneage. To get GLORIA aboard, and in the right position, required more craneage than was available there, and extraordinary efforts were called for from the RVS staff in attendance. In conjunction with the repair yard, they worked out that the only way to do this was to move the ship partway into the drydock, and use the lift of the repair yard's crane to shift GLORIA from the dockside on board. There was, however, one small complication - the drydock entrance was constructed with a sill which, at high water, would provide only about 1.5 metres of water below the keel of RRS *Charles Darwin*, and high water was in the small hours of the morning. Accordingly, at an hour when others slept, the dedicated band of RVS staff trooped down to the dockside, waited for the flooding tide to rise sufficiently to float the ship over the sill of the drydock, then began manoeuvring GLORIA aboard.

Two of the staff watched over this exercise with more than simply technical interest. Ivor Chivers, as Head of Mechanical Engineering, had to ensure that GLORIA was accurately positioned and properly fixed to the after deck of RRS *Charles Darwin*. He quite rightly therefore insisted on the system being carefully lowered on to the deck, and raised and inched into position until he was satisfied that it would hold in a seaway and function properly in use. Paul Stone, at the time Ship Maintenance Manager, but now RVS Marine Superintendent, was concerned for the safety of the ship itself. With only 1.5 metres of water maximum under the keel, he was keeping a keen eye on the dock wall, observing the drying tide mark as the water receded. He knew that if the ship were to ground on the drydock sill, it could break its back, and all would be lost. As Ivor asked for 'Just a few minutes more', Paul fidgeted and began preparations for a rapid reversal! But in the end, this mini-drama resulted in a happy ending, and RRS *Charles Darwin* returned to her berth with no damage and with GLORIA fitted and ready to go.

But the stories did not end there. The ship was berthed at a quay that was both relatively remote and also awkward in shape and size. Two happenings are recalled by support staff sent to prepare the ship for sea. The first is that when the 40 ft refrigerated container of foodstuffs arrived on the quay, the local labourers would not enter its freezing interior to unload it, neither was there a fork-lift truck available. So the RVS staff, who were stripped to the waist because of the 100°F+ temperatures, had to dash into the frozen interior, pull out a pallet of food, which was then lifted aboard by the ship's crane. This was not too much of a problem for the first few pallets, but as the container was unloaded the dash became not only slower because of increasing exhaustion but also colder because of the need to go in further.

The second happening was potentially more serious, but as so often happens it contained within itself a sort of black humour. Some heavy equipment was to be loaded on to RRS *Charles Darwin*, and the RVS staff asked for a crane able to do this: they were assured that one had been ordered. Some little while later, a self-propelled barge was observed approaching. On it was a large, steam-driven crane and a small hut with washing flying from a clothes line. This proved to be the necessary crane, which drew alongside, and hoisted the equipment - a winch - aloft and over the deck of RRS *Charles Darwin*. Sadly, at this critical juncture the steam boiler on the barge decided to cease to function, and the crane's jib and the winch it was carrying descended at a worryingly rapid rate towards the NERC ship's after deck. Disaster was averted by a couple of large, but empty, wooden packing cases which were still on the after deck and which therefore absorbed much of the energy of the descending winch which would otherwise have caused structural damage.

With these problems overcome, the next cruise (CD20/87) could begin, with the formal remit -

> 'To study the sedimentary processes and recent history of the Indus Fan'

to be followed by CD27/87, with a remit

..to conduct sediment sampling, bottom photography and further seismic profiling over that portion of the Indus Fan that was mapped in February 1987, by RRS C*harles Darwin* Cruise CD20/87.

Core samples will be obtained; a WASP Camera system will be utilised, and a single channel seismic system will be deployed utilising airguns and a 3.5KHz sub-bottom profiler. A magnetometer, gravity meter and PES will also be used.

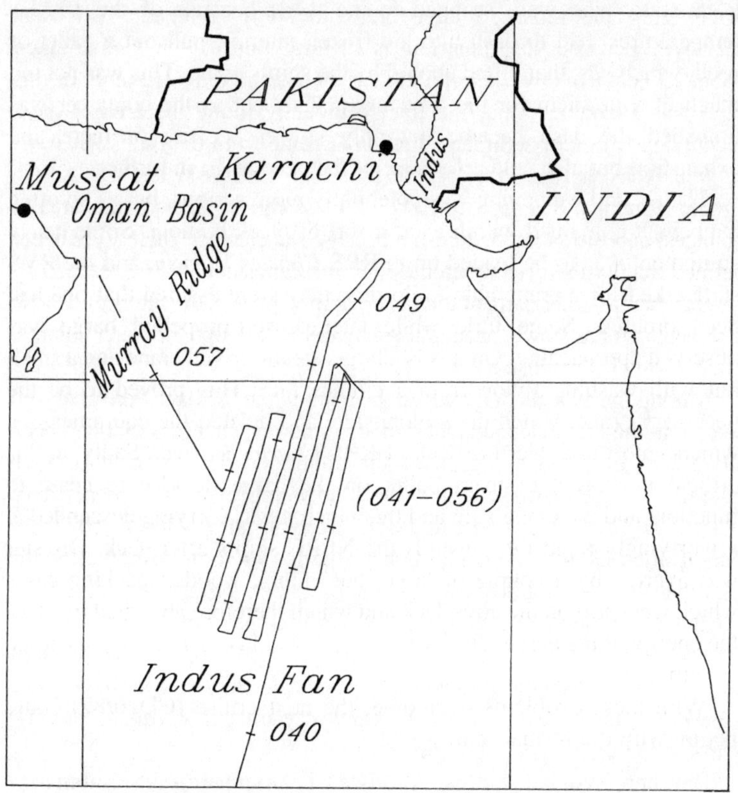

Plate 9 RRS Charles Darwin Cruise 20, 31 January - 27 February 1987

The Indus Fan is one of many similarly-named worldwide phenomena, which are characterised by a vast accumulation of sediment, usually associated with denudations of continents and major fluvial systems, and the two cruises aimed to gather knowledge of the extent and type of the Fan, and of its material composition.

Once GLORIA was aboard, the ship sailed and science began in earnest. However, that bland statement conceals the concern of the first few days when various components of the entire instrumentation system needed the 'bugs' ironed out. Once these problems were resolved satisfactorily by the scientists and technicians on board, a four day reconnaissance across the Indus Fan was undertaken in the hope of discovering the most effective areas to study in detail. This was necessary because the Indus Fan is so huge that it would have taken about 130 days to cover the fan in its entirety, and only one tenth of this time was available. This first line fortunately found the area of most recent deposition of sediments that have come from the Indus River.

Neil Kenyon (the Principal Scientist for CD20/87) and Dr Adrian Cramp (Principal Scientist for CD27/87) had chosen to study this area of youngest deposition on the mid-fan because the channels there were better defined than at the ocean extremes and the deposits could be reached by corers whereas other parts of the fan are buried by tens of metres of pelagic sediment that has slowly accumulated by fallout through the water column. Plate 9 shows the cruise track.

Among the aspects that excited the team when the data were analysed were that they revealed ancient, but still sinuous, channels that had been filled by sand, and evidence of the continued development of such channels, some of those nearer the land being over 300 metres deep (Clark *et al* 1992, Kenyon 1992, Kenyon *et al* 1994). If these conclusions seem to the reader to be esoteric or remote from reality, the fact that they are not is shown by a subsequent enquiry seeking advice about the problems of laying a 1,200 km long gas pipeline expected to cost several billion pounds and transport gas worth some tens of billions of pounds across the Indus Fan. This question would have been answered much less accurately (and certainly less quantitatively) without the data gathered on this cruise.

But more than being a just a study of a particular specific submarine fan, this cruise added to the sum of knowledge of submarine fans generally, allowing a more detailed classification of their types (Clark, Kenyon & Pickering, 1992; Cramp, Rasul & Kenyon, 1994 in press; Kenyon, Ayub & Cramp, 1995; Kenyon, Ayub & Cramp, in press). These papers drew upon a range of data, but the summary to the first cited gives a good indication of the synthesis that can be - or maybe, has to be - developed by scientists working in a global topic area like marine geology.

> Certain attributes of submarine channels measured from GLORIA sidescan sonar data from 16 different submarine fans indicate similarities with fluviatile[18] systems. Channel width, depth, meander radius and wavelength, sinuosity and gradient were measured. This approach makes it possible to identify high-sinuosity, low-gradient (eg Indus Fan channels) and low-sinuosity, high-gradient (eg Porcupine Seabight channels) channel systems as end members. Current classifications of submarine fans relate fan shape to grain size or sediment calibre and therefore are inadequate principally because the shape of the fan is strongly controlled by the shape of the receiving basin, which in turn is dependent upon parameters such as tectonics and diapirism.[19] Overall fan shape is almost invariably independent of the physics of sediment transport. Rather than the fan shape, the geometry and other characteristics of submarine channels and canyons provide a more promising means of differentiating deep-marine turbidite systems.

These cruises produced data which, as in many other branches of science, led to a number of conclusions. The first of these was to add to the sum of existing knowledge - ie that on submarine fans generally. The second was to suggest that the then current theory of submarine fan evolution and classification was not adequate. The third conclusion was the development of a more effective classification of, and hence forecasting method for, the evolution of submarine fans.

Cramp (1987) subsequently reported

> Despite some technical disappointments, Cruise CD27/87 proved to be a great success. In a period of 23 days spent at sea, 35 seabed samples (mostly piston cores) were obtained... We also gathered over 1,000 line km of single channel seismic data.

The GLORIA survey itself was interesting for another reason. The waters of the Indian Ocean are seasonally rich in nutrients, and hence support an abundance of fish life. GLORIA works by sending out energy pulses and recording the reflections from the seabed. When working not far offshore, the scientists were puzzled for a while by signals that appeared to originate from mid-ocean depths, rather than the bottom. After they had worked at night with a light over the ship's side and seen significant numbers of fish at different depths, they concluded that what they were witnessing were echoes from this rich biomass at those depths. As the ship moved into deeper water, the spurious signals disappeared. Unlike the physicists on Cruise CD 22/86, the geophysicists did not view these signals as an opportunity to open up a new field of study.

Peter Miles, a colleague of Neil Kenyon, from IOSDL, had had a long-held interest in seafloor spreading, but knew that 'traditional' geophysical surveys alone would not attract funding, even in exciting areas like the Arabian Sea. When he learned that Neil was to use GLORIA to study the disposition of surface deposits on the Indus Fan he asked him if he would be using other underway geophysical survey methods simultaneously; it appeared not. So Peter asked if room could be found for him to sail on the cruise, and use complementary survey techniques - and Neil agreed. Miles used gravity and magnetic field measuring instruments to gather data to further his studies on the relative movement of the Indian sub-continent away from the Seychelles - something it had done initially at a speed of about 5.5 cm per year until this had been slowed by its contact with Asia. Miles and Roest (1993) used these data not only to show the movement of the sub-continent, but also to provide considerably more detail of the bathymetry of the Indian Ocean in the area than had been known previously.

In his report on this cruise, the Master notes that towing GLORIA along pre-defined, parallel, tracks for 23 days at a constant 8-10 knots, is 'rather boring, with course alterations being a major point of interest'. Since the weather throughout the cruise varied from good to excellent, even this could not provide a talking point. However, when this cruise ended in Mina Qaboos, some improvement in social life took place. There was a delay, for reasons outside the scientists' control, in returning them to UK, and they found Billy Connolly in the Al Falaj Hotel which was used as the preferred base in the city.

Billy Connolly was in the process of making a tour of UK expatriate communities in the Middle East, and the return of RRS *Charles Darwin* to Mina Qaboos happened to coincide with the one night that he would be in Oman. His performance was given in the open air, on a stage erected over the swimming pool at the Hotel. Peter Miles claims that he is unable to recall with clarity any specific jokes from that event, but noted that after a month at sea, he and colleagues '...would have been prepared to laugh at a light bulb'. What he does recall with obvious delight is the joy of ending his cruise on a balmy evening, surrounded by mountains, sharing in the enjoyment provided by one of the UK's premier entertainers. Perhaps there is some divine intervention that balances boredom and stimulation using some rule unfathomable to mere mortals, and the scientists of Cruise CD22/86 qualified as beneficiaries.

Roger Scrutton, from the University of Edinburgh, was next to sail, also on a geophysical cruise. The formal remit of his cruise was

> ..to undertake studies of geophysical deformation in the oceanic crust in the central Indian Ocean, south of the Equator, and between 77°E and 84°E. This is a unique area of compressive deformation that probably arises from collision of India with Asia by Continental drift. Seamounts that may have contributed to triggering deformation will also be studied - including the Afanasy Nikitin Seamount.

What an emotive remit - to study the geophysics arising from the collision of India with Asia! Although that may seem an unlikely scenario, one should remember the fact quoted in a previous chapter, that the tectonic plates on which continents stand are drifting at about 3 cm per year - or put another way, each million years a particular piece of land or seabed could have moved up to 30 kilometres (or just under 20 miles). And as Peter Miles showed in his paper (Miles and Roest, 1993), the Indian sub-continent drifted away from the Seychelles. Thus the collision of the Indian sub-continent with Asia that formed the Himalayas can be dated fairly accurately, and, for instance, changes in the height of the mountains like Everest can be used to estimate the extent of continued movement.

The area being studied on this cruise was some 600 miles - strictly, nautical miles (or nm) - south of Sri Lanka, near the centre of what is known as the Indo-Australian Plate, whose northern boundary is the Himalayas and their continuation into the island chain formed by the Andaman Islands, Sumatra and Java. Hence the formal link between the study area and the Indian -Asian collision.

Much of the primary interest of modern geophysics has centred on the boundaries of the tectonic plates - see the map on Colour plate 2. However, the centres of the plates, particularly those under the oceans, are also of scientific interest, and Scrutton had chosen to study this particular one. His expanded scientific objectives (Scrutton 1988) were:

i) To collect multi-channel seismic, single channel seismic and sediment velocity data in the vicinity of ODP Leg 116 sites. These were to contribute to an understanding of the structural setting of the drill sites and will be worked up for publication in the Leg 116 Proceedings volume.
ii) To study the three-dimensional form of the sediment and basement deformation in the area of folding and faulting around 5°S, 8°E. Single and multi-channel profiling and disposable sonobuoy data were to be collected. Gravity and magnetic data were also to be acquired to contribute to an understanding of crustal thickness and tectonic setting. A structural analysis will follow from this.
iii) To sample basement and sediments on Afanasy Nikitin Seamount and collect enough magnetic, gravity and seismic profile data to allow an integrated geological -geophysical study of the seamount. It is hoped to date the formation of the seamount, which may have provided a critical load on the lithosphere to promote deformation.

In fact, because by this time the ship had considerable fouling below the water line and her cruising speed was limited to 9.5 knots, instead of the 11 on which the programme had been postulated, and because it became necessary to make a portcall at Colombo to collect a spare part for one of the compressors on board, objective (i) was abandoned. Such are the vagaries and perils of marine science.

However, once in the working area, and with the equipment functioning well under the tender ministrations of the support technicians from RVS, Barry, research began in earnest. Scrutton reported 'The work carried out at sea on this cruise was a success, and the scientific objectives have been met, if not completely satisfied'. This success was to show that the intraplate deformation in the area could be attributed to a re-activation of the forces that caused the original structural faults (Bull 1990, Bull & Scrutton 1992.) and could be dated at about 7 Ma (ie 7 millenia before present day). In addition, further analytical and analogue studies by Bull and others (Bull, Martinod and Davy, 1992; Beekman, Bull, Cloetingh & Scrutton, in press 1994) deduced the mechanisms implicit in the faulting and folding revealed by the data gathered on the cruise.

This cruise ended in Mauritius , where the ship's passage crossed that of HMS *Beagle* for the first of many such crossings. HMS *Beagle* was nearing the end of her voyage having, as noted early in this narrative, gone 'west-about' rather than the easterly passage of RRS *Charles Darwin.* By January 1836, just over four years after sailing from Plymouth, even the intellectual curiosity of Charles Darwin was becoming jaded and his homesickness more apparent. He notes

> All which we have yet seen is very pleasing. The scenery cannot boast the charms of Tahiti & still less of the grand luxuriance of Brazil; but yet it is a complete and very beautiful picture. But there is no country which now has any attractions for us, without it is seen right astern, & the more distant & indistinct the better.

RRS *Charles Darwin* had, on arrival in Mauritius, been away from the UK for about 18 months but, unlike HMS *Beagle*, she had not been manned by a single crew and scientific party for all of that time. For each new cruise, crew, port and scientific team there were new stimuli to maintain interest, and although many of the officers and crew and, to a lesser extent, the scientists themselves spent extended periods on the ship and visited a number of locations, none of the reports resulting from this circumnavigation express boredom or such a deep longing to return to the UK as the great man himself recorded. Thus whilst HMS *Beagle* continued her voyage by sailing round the Cape of Good Hope,

RRS *Charles Darwin's* contact with South Africa was at Durban, thence to continue her eastward peregrinations.

---

[17] The reference is to Arthur Fisher, who was the long-time logistics officer at IOSDL

[18] fluviatile - of, found in, or produced by rivers

[19] diapir - a dome or mushroom shape often assumed by a body of light material that is rising through the crust because of its comparative buoyancy

6
The Long Transect

RRS *Charles Darwin* was now in Durban, South Africa, on Longitude 31°E. Almost 18 months after leaving home, the ship had completed less than 10% (in terms of longitude) of her projected circumnavigation. What to do? Ship time is expensive, passage time a waste scientifically. Enter the American cavalry, ably led by John Toole from the Woods Hole Oceanographic Institution, not quite wearing stetsons and riding white horses, but certainly keen to survey the white horses of the Southern Indian Ocean.

The formal cruise requirement was to:

> ..investigate aspects of the Southern Indian Ocean. This will involve a transect, west to east, between latitudes 29°S and 35°S, consisting of 125 full ocean depth casts to monitor conductivity , temperature, depth, and dissolved oxygen.
>
> Characteristics of various core water masses; the zonal extent of the subtropical gyre; the structure and transport of deep western boundary currents; the relationship between Australian Coastal circulation and deep ocean flow; the net Meridional fluxes of heat and fresh water into the Indian Ocean across the section are all aspects under study.

And so, in mid-November 1988, with the wide Indian Ocean stretching beckoning and unconstricted clear across to Australia, RRS

Charles Darwin prepared to sail from Durban. But the dreaded logistics bug struck again, and the scheduled departure date arrived, but not the airfreight from the University of Miami! Luckily, the delay was only a matter of hours, rather than days, and the ship left Durban at 2100 on 12 November. Even now, the siren call of the wide ocean had to be resisted for a while, because, less than 24 hours after departure, a family emergency required the ship to return to Durban to disembark one member of the crew and pick up his replacement. Because the replacement member of the crew would not be in Durban until pm on 15 November and because the wind had risen to 40 knots, John Toole, the Principal Scientist, decided to occupy some coastal stations on the way in and to effect some equipment tests. And so, it was not until 1700 on 15 November that Captain Sam Mayl was able to point RRS *Charles Darwin's* bow due eastward and leave it there for some 39 days, the longest cruise so far and close to the ship's maximum endurance of 42 days. And, as he noted in his report, the Master was proud of the fact

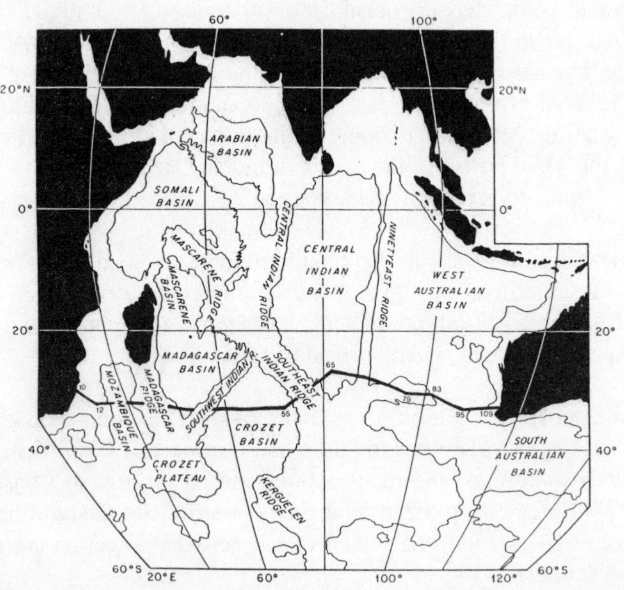

Plate 10: The trans-Indian Ocean cruise track and CTD station locations of RRS Charles Darwin cruise#29 from Africa to Australia. Note the many ridges and basins traversed by the cruise track.

RRS *Charles Darwin* passing through the Thames Barrier, courtesy NERC.

EURASIAN PLATE
NORTH AMERICAN PLATE
PACIFIC PLATE
EURASIAN PLATE
AFRICAN PLATE
SOUTH AMERICAN PLATE
NAZCA PLATE
INDO-AUSTRALIAN PLATE
PACIFIC PLATE
INDO-AUSTRALIAN PLATE
ANTARCTIC PLATE
ANTARCTIC PLATE
ANTARCTIC PLATE
NOTE
DISTRIBUTION OF VOLCANOES AND EARTHQUAKES
INTRODUCTION
EXPLANATION
ACKNOWLEDGMENTS
Schematic cross section illustrating plate tectonics processes
PREPARATION OF THE MAP
SELECTED REFERENCES
THIS DYNAMIC PLANET
WORLD MAP OF VOLCANOES, EARTHQUAKES, AND PLATE TECTONICS

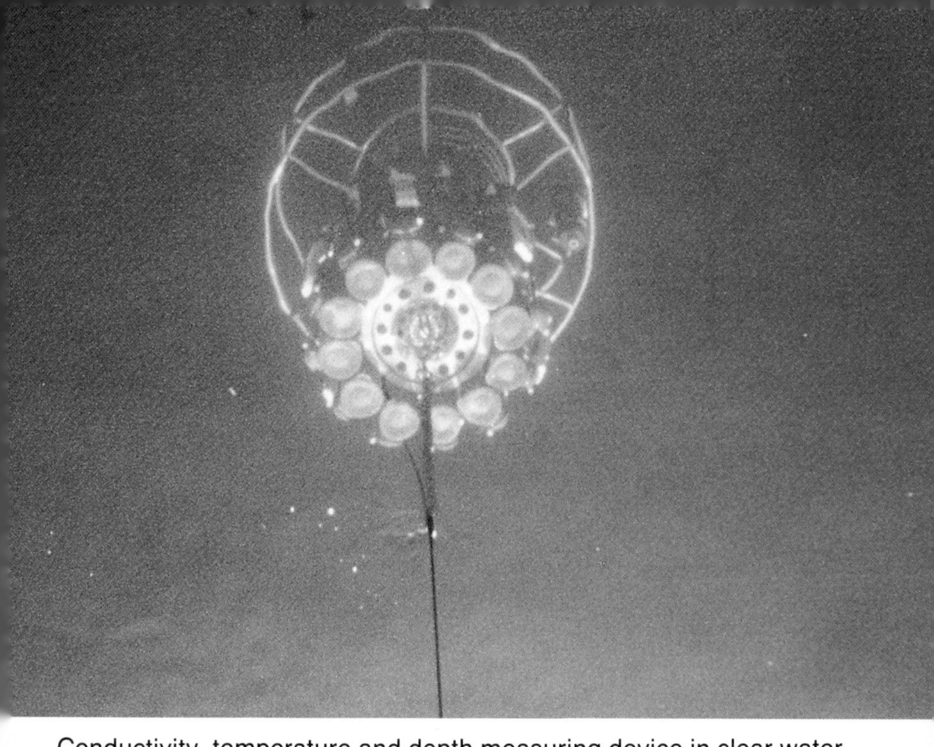

Conductivity, temperature and depth measuring device in clear water, courtesy Captain K. O. Avery.

Core sample, courtesy Dr A. Cramp.

Fine view over stern, courtesy Captain K. O. Avery.

Barbeque, courtesy Dr M. Woodward.

Rough view over bow, courtesy R. Larter, British Antarctic Survey.

Visitors on Aldabra, courtesy Dr K. Heywood.

Crossing the line,
courtesy Professor D. Olson.

Gloria deployment, courtesy R. Larter.

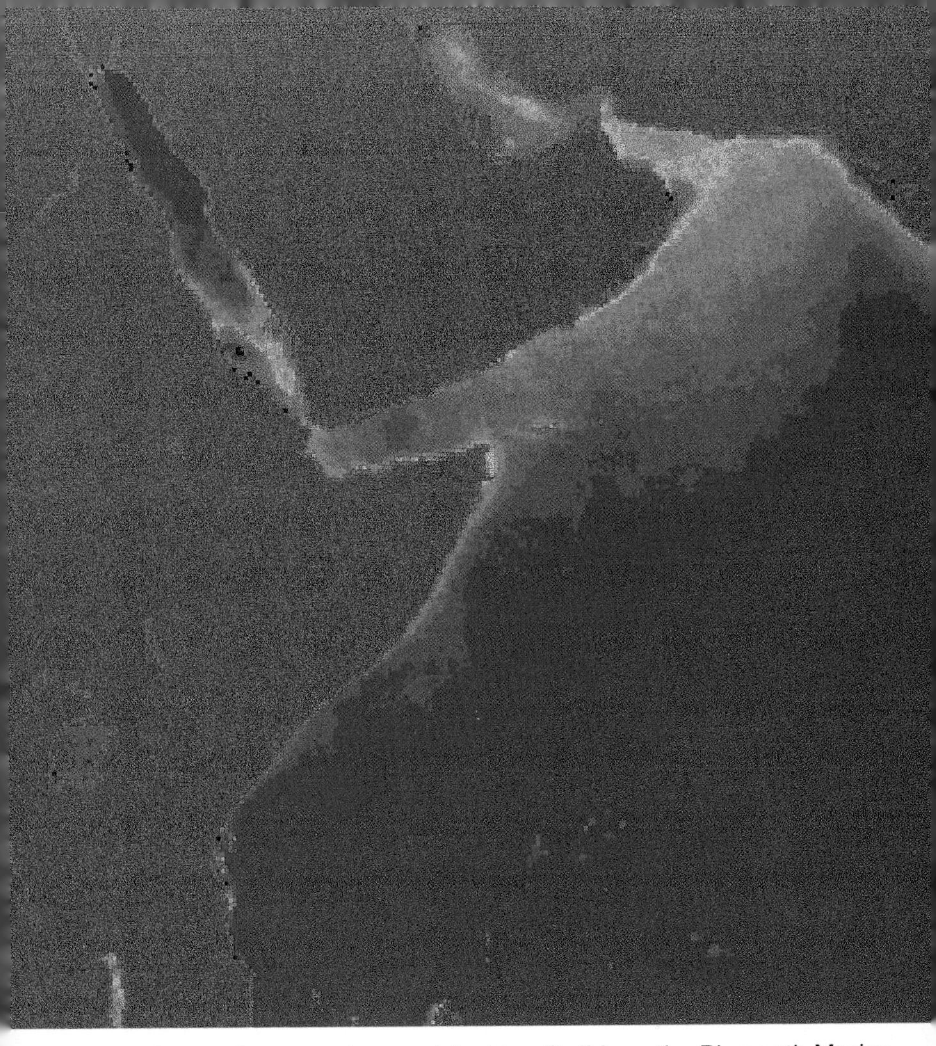

oastal Zone Colour Scanner picture of Arabian Gulf from the Plymouth Marine Laboratory cruise.

Committal ceremony of the ashes of Captain Sam Mayl, courtesy Dr S. Tarbell.

RRS *Charles Darwin* leaving Woods Hole Oceanographic Institute, courtesy Dr S. Tarbell.

that this cruise marked the return of a British Research Ship to Eastern Waters after more than 100 years.

But there was another comparison that could be made - Charles Darwin completed his circumnavigation in HMS *Beagle* by sailing from Sydney westwards in a less direct manner than his later namesake. The naturalist had been away from England for just over four years at the time he left Sydney (January 1836), but it was a further five months before HMS *Beagle* reached Cape Town. His journey was not then to be directly homewards because, as he notes in a letter to his sister:

> Some singular disagreements in the longitudes made Capt F R (Fitzroy) anxious to complete the circle in the southern hemisphere, & then retrace our steps by our first line to England. This zig-zag manner of proceeding is very grievous; it has put the finishing stroke to my feelings. I loathe, I abhor the sea, & all ships which sail on it.

The weather by now had improved, and was good for the remainder of the long voyage of RRS *Charles Darwin*. Indeed, it was so good that Professor Rana Fine, the chemical oceanographer on board, actually claimed to have enjoyed the trip. What is unusual about that statement is that Rana had supervised the design by Kevin Sullivan of the techniques that were used on the earlier American-led cruises in the Arabian Sea. However, she had declined to sail on those because she 'suffered unduly from the Charles Darwin syndrome, aka sea-sickness'. On this cruise, she was able to analyse the data and even enjoy doing so, while Kevin Sullivan and his wife Leslie Pope did the chemical analyses. They had also done the analyses on the other two American-led cruises, and they, Bruce Warren and Bill Miller, an RVS technician, were the only four to have participated in all three.

A routine now began to emerge. Sail to a selected station, take a CTD cast; when the CTD was inboard, set sail for the next station. On board the scientists would begin analysis of the data provided by the CTD, or of the water samples gathered at the required depths, perhaps streaming an XBT probe to monitor the primary water parameters during the passage. The shipboard technicians from RVS would maintain the equipment to minimise the possibility of data loss, and to eliminate as far as humanly possible the prospect of losing equipment. Only two matters were slight cause for concern in this seemingly idyllic setting. The first was that the ship, now some 12 months since her last refit in Mombasa, was being slowed by growths on her bottom. Her design cruising speed of 12 knots had become a practical maximum - 'with a following wind ..'. as her Master reported -' of 9 knots'. Proof of this growth, if indeed proof were needed, came when Sam Mayl flew back to UK from Singapore with a polyfoam cup containing, but only just, a huge barnacle with weed attached! The second problem was that the combination of large equipment and relatively slow winch speed meant that each station took longer than had been projected. However, as time passed, crew, scientists and technicians worked together to improve the performance of all the equipment, so that at the end of the cruise John Toole could report - 'The data return from the cruise was exceptional; the major cruise objectives were met. This may be credited to careful preparation for the cruise, and hard work by both scientific and shipboard personnel during the trip'.

So what were the objectives of the cruise in terms less formal than set out above? The Indian Ocean is bounded on its westerly side by the continent of Africa, and on its easterly less rigidly by the island chain that begins with the Malay Peninsula in the north and runs down through Sumatra and Java to Australia. In the south the Southern Ocean forms a boundary between the Indian Ocean and the landmass of Antarctica.

In the early 1980s, an increasing number of scientists came to recognise the interaction between the oceans and the atmosphere, and the fundamental ways in which the former affected the latter at the meteorological and climatic levels. Also, in parallel, increasing significance was attached to the global nature of any studies of this topic, and ideas began to form about international programmes. One of these eventually became the World Ocean Circulation Experiment (WOCE), and this Indian Ocean Transect was either a precursor to WOCE or one of the first formal WOCE experiments, depending upon which source of information is consulted. The specific interests of the cruise were, therefore, to study the internal circulation of waters in the Southern Indian Ocean.

The basic techniques of the study used were the CTD and ADCP systems spelt out in earlier chapters. Another technique (as on the earlier American-led cruises in the Gulf of Arabia) was the chemical analysis of water samples for the concentration of dissolved Chloro-Fluoro-Carbons (CFC). CFCs are a relatively recent invention, as they were designed, *inter alia*, to replace hydro-carbons in propellant sprays, such as hair lacquers, and heat-exchange fluids in refrigeration and air-conditioning plant. One of their properties which was initially regarded as a virtue was that they did not break down rapidly: subsequent research has shown this to be an atmospheric problem, and research by Farman[20] laid the blame for the depletion of the ozone layer over Antarctica directly at the door of CFCs. However, their stability does provide a useful monitor for oceanography, since the concentration of specific CFCs in seawater will give a measure of the age of that watermass, and hence an indication of its origins. In crude terms, the older bottom water will be free of dissolved CFCs, but if there is a measurable CFC concentration it can be used to show how long since the water was last in contact with the atmosphere. During this long transect, Rana Fine found the first low-level concentrations of CFCs entering the bottom water of the subtropical Indian Ocean in the Crozet Basin about 60°E. She estimated their age at less than 15 years since leaving the probable source region of the Weddell Sea.

So what did this cruise reveal from a combination of these measurements? To quote Toole and Warren (1993)

> ....water property features in the upper kilometre indicate that the northward interior flow is predominantly in the eastern half of the ocean there (the southern part of the South Indian subtropical gyre), consistent with the forcing pattern of wind-stress curl. The southward return flow is the Agulhas Current.......Circumpolar Deep Water flows northward to fill the greater deep Indian Ocean by means of western-boundary currents in the Crozet Basin, Central Indian Basin and Perth Basin. North Atlantic Deep Water entering directly from the mid-latitude South Atlantic is almost entirely confined to the south-western Indian Ocean...by the topography of the Madagascar Ridge and the Mozambique Channel.

These summarisations were, of course, subject to qualification in the body of the paper referenced, but only in quantitative terms.

Toole and Warren (1993) conclude their paper with..

> The present study finds that within the Indian Ocean north of 32°S, deep and bottom waters are converted largely to thermocline and surface waters. This process inherently involves mixing; one speculation, here based on hypsography[21], is that mixing processes near boundaries might be particularly important in the Indian Ocean. Upwelling, of course, is the counterpart to deep convection which occurs in high-latitude sites in the Atlantic and Southern Ocean, At those sites surface waters are converted into dense deep and bottom waters. Despite its small size, the rate of deep-water conversion to thermocline and surface waters in the Indian Ocean appears to be large. The Indian Ocean thus plays a significant role in the global thermohaline circulation.

Fine (1993) used CFC tracer analyses to study the circulation of Antarctic Intermediate Water, which is formed by mixing either across the polar front in the southwest Atlantic or north of the Subantarctic front in the southeast Pacific. Antarctic Intermediate Water is transported eastward by the Antarctic Circumpolar Current, and penetrates into all three northern hemisphere oceans - Atlantic, Pacific, and Indian - and Fine wanted to examine its circulation in the Indian Ocean and compare it with that in the other two oceans. Her summarised conclusions as regards Antarctic Intermediate Water were four, viz:

> 1 The most recently ventilated Antarctic Intermediate Water is observed in a compact anticyclonic gyre west of 72°E. The shallow topography may deflect and limit the eastward extent of the most recently ventilated Antarctic Intermediate Water.
>
> 2 The Agulhas Current may impede input of Antarctic Intermediate Water along the western boundary.
>
> 3 Tracers are consistent with an inter-ocean flow from the South Pacific into the Eastern Indian Ocean, similar to the South Atlantic to Indian linkage.
>
> 4 It appears that the high wind stress curl forces an equatorward component of the circulation that is strongest around 60°E.

Whilst it is undoubtedly true that each of the above conclusions represented, in its way, a scientific step forward, a contribution to man's understanding of the complexities of the oceans in general and the Indian Ocean in particular, there is no doubt which Rana Fine was most excited about. Conclusion 3 above - that there is an exchange of water between the South Pacific and Eastern Indian Ocean - is not simply an addition to existing knowledge, but something totally unexpected. It was possible to detect, indeed impossible not to, when talking to her that being a principal participant in unveiling this discovery was one of the inherent reasons for becoming a scientist in the first place.

So much for the science, but what of the people? It is first worth recording here a theme that recurred, of how effective RRS *Charles Darwin* was (and, indeed, at time of writing still is) as a research platform. John Toole records in his cruise report

> The *Darwin* (*sic*) is one of the best ships I have worked from. A positive attitude exists aboard and all ship's personnel are eager to help with the science. ...I wish to acknowledge the four RVS technicians who went beyond their nominal duties to assist us in our work.....Our thanks are extended especially (to the catering department) for our Thanksgiving Day dinner and the cookouts we enjoyed during the trip.

Therein, perhaps, lies some of the secret of the way science gets done. Not only a purpose-designed platform and equipment, but close co-operation between scientists and supporting personnel, whether crew or technicians. After all, Thanksgiving Day is an American holiday, so what possible reason could there be for a British crew on a British ship in the middle of the Indian Ocean to make any effort to celebrate it? Apart, that is, from encouraging the already friendly relationship developing with the American scientists on board. What the crew may have had in mind was that Thanksgiving Day, as defined by Webster, is..'.set apart in the USA for recalling the goodness of God in blessing the Pilgrims with their first good harvest (1621)'. And, of course, those pilgrims were British, having set out from Plymouth, the very port from which Charles Darwin himself set forth. So perhaps the catering department felt they should continue a tradition started by their forebears.

But there could be more to it than that - the ship's newspaper of the time, *The Beagle Times*, carried an 'Agony Aunt's page' (when a particular RVS technician was aboard to provide the advice!). Towards the end of this voyage, a letter appeared, signed 'MF'. This letter read:

Dear Aunty

I had hoped to persevere through this entire cruise without having to consult you or your column. After all, I've been to sea several times and take some pride in leaving a ship on good terms with everybody on board. However, with numerous differences surfacing between British and American ways and customs, I was hoping you might lend me some pointers to help me through the short time left at sea without any major international blunders.

What follows is a sort of 'how I see it'. I'd hate to jump to conclusions and make judgements on misinformation, so please ... correct me if I'm wrong... you always do.

As I understand it, meals are a highlight of British social activity. Really, it can be quite the same in America. Of course, rumor has it that we have a much more diverse diet and socialize constantly rather than at meals, but we're basically the same. However, a few small things do strike me as different from the good old 'home on the range' style of living.

On the *Darwin*, each meal sitting consists of multiple courses with just the appropriate level of harassment to keep everybody on their toes. I've learned (after numerous mistakes) never to put my knife and fork together if I have any intention of having another bite. Also, in America, we have a custom after a good meal of simply commending the cook. Is it true what I've been told, that eating excessive amounts of haggots (or faggots, I can't remember which) signifies the highest level of appreciation to the English cook?

I've found that the science party here always eats in the officers' mess. This goes along somewhat with the preconceived notions I arrived with: that British folk are excessively organised if not highly regimented (nothing personal, I picked this up sometime during childhood from a father whose first two names are Samuel Adams). Anyway, those notions have perished, and could even perhaps be taken on by the Americans. I find that no one, thank goodness, has assigned seats (except the Captain), yet everyone (except the Captain) sits in the same seat every meal. Each person's place at table is identified by their 'soup signature' engrained in the tablecloth. As custom will have it, all stains are pointed out to fellow eaters by the steward. Some are larger than others. I believe Bruce Warren's seat has the most variety.

Aside from meals, though, Brits and Americans tend to be extraordinarily similar. Not surprizing, I guess.

During daylight hours, ship's watches consist of SOME time on the bridge, and LOTS of time at the shuffle and dart boards. Science watches consist of SOME time in the lab, MORE time on the bridge, and LOTS of time at the beach.

Both parties spend the rest of the hours eating and/or drinking.

Tanning competitions seem to be the ritual among both nationalities. The British use the engineers' jump suits as the standard white 'blank' for comparison. The Americans use the PSO. Much to poor George's dismay, nobody told him that one of the first rules of either team was 'you peel, you lose'.

Language, luckily, doesn't seem to be a problem among us. Something in my favor anyway ... Chinese ships are a bit tricky at first, but make this seem like a piece of cake (pudding). After all, in American or British, a scaaaaf is a scaaaaf, and a daaaart is a daaaart. In New England, where I come from, we know that there are others less fortunate in the country who talk with funny accents so we accept them as just a little odd ... the same as those from England accept you from Wales. No hard feelings, I'm sure. It's obvious, all participants from all sides enjoy a good barbeque now and again. A dart game, a shuffle game... always in competition between the Brits and the Good Guys, but all in good fun. We even mixed teams a little this last round. Obviously the Brits decided too many Americans on one team would be more than they could handle.

Anyway, Aunty, I hope I make it through. I'd hate to make a mark in history between the US and England. I'm not the ALL AMERICAN I'm supposed to be, but I'm taking lessons during watch and practicing on my own to save the fate of diplomacy. Maybe you can help.

The short answer from the Agony Aunt doesn't add to this cultural interchange, and might be thought to lower the tone of this book! However, MF (assumed to be Ms M Francis, from the Woods Hole Oceanographic Institution) has made her contribution to Anglo-American harmony, and provided evidence of the way that working and personal relationships between colleagues can come to terms with differing cultures. With, as noted, an increasing emphasis on international research programmes coupled with national pressures on the funding of science, this ability to work together both personally and on shared facilities will assume added significance.

---

[20]Farman J, *Depletion of the ozone layer above Antarctica, Nature* Vol **315** (16 May 1985):6016.

[21]A branch of geography that deals with the measurement and mapping of varying elevations of the earth's surface (Webster).

7
Singapore Refit

The end of 1987 marked the mid-point of the circumnavigation, but it was also a year since RRS *Charles Darwin* had left Mombasa after her refit. It was time now for a second annual refit, and not simply on the basis of time elapsed. In the previous chapter we noted that growth (or growths) caused by the warm Indian Ocean waters below the ship's waterline had been seriously reducing her speed. A consequential effect was an increase in fuel consumption - Sam Mayl noted that when fresh out of Mombasa, the fuel consumption had been approximately 5.9 tonnes per day, whereas the long transect had consumed 7.8 tonnes per day, an increase of over 30%. Clearly, NERC could not afford to continue the voyage without taking remedial steps, and so a refit was scheduled.

Following the normal tendering exercise among shipyards in the general geographical area able to carry out the work, a decision was made to effect the refit in Singapore. But the ship was in Fremantle, in Western Australia, some 2,000 nm away, and an unused passage was anathema to scientists. And so an extra cruise was scheduled, on the basis that any science was better than no science. Scientists from the University of Liverpool Department of Oceanography had long been interested in air particulate sampling - ie gathering dust samples - and they leapt at the opportunity to use the voyage from Fremantle to Singapore to collect samples from a part of the world that they would not normally be able to visit.

And so, at 1320 on 19 December 1987, RRS *Charles Darwin* left Australia and set sail generally nor-nor-east for Singapore. On

Christmas Eve, the ship passed Christmas Island, but no written record exists of what went on that night or the following. Doubtless, those crew and scientists not on watch enjoyed the odd glass or two of refreshment (tea, of course!), and it is likely that the satellite communications channel was well used as greetings were exchanged across the world. The Master may well have drawn comparisons with Christmas 1986, where villainy was abroad and murder stalked the alleyways and showers, but perhaps not. His record only notes drily 'I berthed at Keppel Shipyard, Singapore, at 1615 hours on 28th December'.

RRS *Charles Darwin*, at a mere 70m, was dwarfed not only by the drydock into which she was put but also by the sheer scale of the bulk carriers, tankers and cattle ships sharing the yard. Surrendered to the ministrations of shipwrights, fitters, welders, and painters, but watched over by specialists and engineers newly arrived from Barry, the ship ceased all scientific activity for a month. She submitted to the indignity of having her bottom scraped and repainted, her engines went through the land-based vehicular equivalent of a 'de-coke', and various other checks and balances required by the certification authorities were carried out. To some of those from UK watching over the ship, the techniques bordered on the archaic - some of the hull painting, for instance, was effected by men on the dock bottom using brushes on the end of long bamboo poles. However, perhaps this approach accorded with the view of an officer no longer with NERC, who is reported to have commented that ..'all of this technology is very well, but what this ship needs is a really good plumber!'

But it was not just the techniques that appeared to be from a bygone age - the manning practices also seemed to use as their role model those halcyon days when the senior staff in UK shipyards could be readily distinguished by their bowler hats. Whilst the Singaporean managers did not wear bowlers, they were from an upper class whose principal means of communication was Mandarin, a language not normally understood by the workers. There was therefore an intermediate managerial grade which interpreted the Mandarin commands into language that the workers could understand, and one of these middle managers was assigned to be available to the RVS staff attending RRS *Charles Darwin*. This manager conveyed very clearly the ethos of the yard when an unfortunate accident occurred.

When in drydock, the ship's Chief Engineer wanted to check some crucial measurements on the propulsion system, and decided to turn the propeller shaft by hand. Unknown to him, some scaffolding had been erected around the propeller, and one plank was closer to one of the propeller blades than it should have been. When the propeller was turned, the centre of this plank was displaced, and as the propeller continued to turn it released pressure on the plank, which sprang back, throwing one of the yard workers into the dock. He was slightly injured, but when Paul Stone asked the yard manager who he was, and how he was, the manager said he didn't know the man's name, only his yard number: apparently, all the workforce was known by number only.

Previous cruises had made use of the ADCP which, being fitted under the hull, was not immune to the growth and corrosion affecting the ship's bottom. Sadly, because it is a delicate instrument, it could not be scraped and painted at the same time as the hull. It was therefore removed when the ship entered drydock and returned to its manufacturer for overhaul. It was not returned until the ship had left drydock, and it had to be replaced in its mounting by divers, an interesting but in the end relatively uneventful procedure. Whilst all this was going on, the ship attracted media attention, and Ivor Chivers (the Chief Mechanical Engineer, from RVS) gave an interview on the Singapore equivalent of the BBC's '*Today*' programme. At end-January 1988, RRS *Charles Darwin* was pronounced 'fit for duty' again, and ready for yet more scientific endeavour.

Whilst all this was going on, one or two of the RVS staff present used their free time to explore Singapore. Just across from the shipyard in which RRS *Charles Darwin* was being refitted was an island, which could be visited either by ferry or by an overhead wire ropeway. Four of the staff decided to use the latter, which was some considerable height above the sea in order to allow unimpeded passage to large ships using the docks and shipyard. The four who intended visiting the island by this means were, to use current terminology, 'gravimetrically challenged' - ie all of them were above average weight. They all climbed in to one of the ropeway cars, and on checking the notice which set the weight limit, observed that they were pushing it a little. When the car set off, they were somewhat disconcerted that when it left the station and began the long traverse over the sea, it appeared to drop alarmingly, so much so that they decided to return by ferry.

Shortly afterwards the ropeway was closed for servicing, in the course of which the wires themselves were replaced. This encouraged a couple of the previous travellers to make the journey across the estuary again by ropeway, and this time it seemed to be uneventful. When they reached the island station, one of them - a mechanical engineer by profession - noticed a pile of the replaced wires, which had been cut into short lengths prior to disposal. He went to inspect this a little more closely, and voiced a terse expletive of horror: one of the lengths of wire crumbled to rust dust in his hand, and shortly before its replacement he had trusted his life to its safety! Another of the company commented more pragmatically that the ropeway's engineers had obviously a nice sense of timing in setting refit schedules.

The ship's next cruise was to be off Indonesia, and it was planned to hold a reception and exhibition on board to show the authorities there what science was involved. Scientists boarded in Singapore, and began the installation of their equipment, principally GLORIA, and the ship left the dockside ready to go. However, as the time of departure approached, concern grew that formal clearance had not been given by the Indonesian authorities to work in their waters, and so RRS *Charles Darwin* anchored in Singapore Roads awaiting the necessary piece of paper. The satellite communication channels between Singapore and Barry and Jakarta ran hot with question and answer, until some twenty four hours after the original estimated time of departure word was received that the required approval had been given, and the ship set sail.

8
Southwest Pacific

On passage from Singapore to Tanjung Priok (the port for Jakarta) the scientists aboard were divided into watches and kept 'pirate watch' because the area had achieved a certain notoriety in that respect. The Master also instituted additional radar watches to keep an eye out for vessels that might, for whatever reasons, be sailing without lights, but he refused permission to mount a wooden silhouette of a machine-gun that had thoughtfully been provided by person or persons unknown. Fortunately nothing more exciting than seasnakes shimmering on the sea surface was observed, and the passage was uneventful. In between the watches, the Main Laboratory was laid out with exhibition material and readied for the reception to be held on the ship. The principal aim of the exhibition was to display NERC's scientific credentials and competence.

On arrival in Indonesia, a wide spectrum of officials and expatriates was invited to the reception. The general perception afterwards was that both exhibition and reception had achieved their aims, with considerable social interaction between scientists, crew and visitors. Whether this perception was enhanced or not by an exhibition of 'private enterprise' by one of the junior officers is not recorded. This particular individual, observing that within sight of the ship was a house occupied by several 'ladies', invited a number of them aboard. In the words of one of the scientists present at the reception:

Late in the evening, while there were still many distinguished guests being hobnobbed with, the doors to the laboratory burst open and a certain member of the ship's company arrived arm-in-arm with two ravishing ladies of the night, all set for his own version of a party. Unfazed by the presence of the array of top brass that confronted him, said gentleman proceeded to pass among the assembled group introducing his 'friends' with gusto...no further recollections are possible!

The marine geophysics of the area around the Indonesian Islands had long been of interest to UK scientists, but two things had precluded research in the area. The first was that it was not an area which UK ships visited often. The second was touched upon in an earlier chapter, namely that diplomatic clearance to work in their waters was not readily given by the Indonesian authorities. By overcoming both of these obstacles, the combined efforts of all the NERC and FCO staff involved had opened a door on to a potentially very fruitful scientific area.

And so, after the party, science began again in earnest, led by Doug Masson from IOSDL. Doug, a curly-haired, quietly spoken Scot has developed an interest in the geophysics of the area around Indonesia, and had been co-author with Audley-Charles of the proposal to NERC to carry out the research in the area. When Professor Audley-Charles, who worked with Chris Adams in Indonesia to obtain the formal clearance, was unable to act as Principal Scientist for this cruise, Doug seamlessly assumed the role of Principal Scientist with Dr John Milsom of University College London as co-chief.

The formal remit of this cruise was:

to obtain GLORIA, single-channel seismic, 3.5kHz gravity, and magnetic data over two areas of complex tectonics in Eastern Indonesia.

In the first area, at the eastern end of the Timor Trough and the Savu Basin, an area where the Australian Continental margin is overriding the Java Trench subduction zone will be examined.

In the second area, to the north-west of Irian Java, a little known area where the Philippine, Yap-Palau and New Guinea Trenches come together will be examined.

Strangely enough, for once Masson's formal description of his scientific aims matches those set out in the Sailing Instructions precisely. The areas identified excited scientific interest in part because there was little data available on them and in part because ODP had plans to drill at two sites within this general area and required further data on which to base a final decision.

The cruise began, as had almost become the norm, with concern about the functioning of the equipment. The bulk of the supporting systems for GLORIA had been on board the ship, in a non-air-conditioned container, for some months, and when this was opened it was discovered that the internal temperature had, at some stage, risen to beyond 50°C as revealed by a mercury thermometer which had exploded. These conditions had caused some equipment corrosion, and other problems which had to be overcome before GLORIA could be deployed. In parallel, a malfunction of the ship's engines also caused a 12 hour hiatus. Fortunately, as Masson reports, sufficient systems were repaired to allow an effective cruise, but there was much nail-biting as there were then no spares available for some crucial components.

Reporting on the first aim of the cruise, Nichols *et al* (1990) note 'The objectives of this present study were to define more precisely the position and form of the southern termination of the Philippine Trench and to determine which of the plate boundary models is most applicable'. This aim arose because a number of researchers, between 1979 and 1987, had developed quasi-theoretical models of the Philippine Trench based on limited geophysical data. The data gathered on this cruise enabled Nichols and his co-workers to dispute some of the earlier hypotheses and put forward another, more quantitatively-based, model. For this model, Nichols *et al* (1990) argue that the Philippine Trench displays the characteristics of a subduction zone (ie one of the tectonics plates is riding over the other) which could be expected to expand southward but is constrained at its southern end by the East Morotai Plateau (at approximately 2° N 129° E) , where the ocean floor shelves steeply from over 5,000 metres in the Trench to only some 2,000 metres on the plateau.

On the second aim of the cruise, Masson *et al* (1990) report on the effect of subduction on seamounts on the Java Trench. A 'seamount' scarcely needs definition, but it is a significant protrusion above the surrounding seafloor: one such imaged by GLORIA was some 1,500m high, with very steep sides, and others were observed with base diameters of from 10 km to 60 km. To understand something of the mechanism being studied it is necessary to consider in an elementary way the basic mechanics of subduction.

The areas of the earth's surface that constitute the major and minor tectonic plates may, in general terms, be considered to be inflexible and of finite thickness. When two of these plates collide, relative movement between them involves one of the plates being 'subducted' under the other. Because the lower plate now has to curve to make the transition, its upper surface will be in tension, and this surface is likely to exhibit cracks or faults - an analogy would be with a metal plate whose upper surface dulls and crazes if bent in a radius that is small relative to its thickness. Now imagine that on the upper surface of the subducted plate there is a major protrusion (a seamount) - what happens to the seamount as subduction proceeds? Is it eroded by the relative motion? Is one side eroded, so that it 'falls over'? Is it pushed in the direction of the subduction? What impact does it make on the upper plate? These were the sort of questions that this part of the cruise aimed to address.

Masson *et al* (1990) conclude 'Our data confirm that seamounts being carried into the trench are fractured on approximately the same scale as the surrounding ocean floor'. In other words, the seamount is ridden over by the upper plate as though that plate were flexible, rather than rigid. This is, perhaps, an over-simplification because it implies that the upper plate flows over the lower, seamounts and all, like water, and that is not so. A better analogy would be with a diaphragm, whose horizontal and vertical dimensions sum to a constant: as it flows over the seamount, the vertical displacement increases to adapt to the new contours, resulting in a reduction in the horizontal dimension. Thus relative to some fixed point, the nose of the upper plate will not be as far forward as it would have been had there been no seamount to distort it.

The areas covered by this cruise took the ship south westwards from Indonesia towards Australia again, and RRS *Charles Darwin* arrived at Port Darwin, in the Northern Territories, on 6 March 1988. The port was named after the famous naturalist, but HMS *Beagle* did not call there when Charles Darwin was aboard. It was on a subsequent visit that Lieutenant Stokes named it in honour of his good friend, whom he had accompanied around the world. The visit by RRS *Charles Darwin* coincided with the 150th anniversary of the naming of the port, and was widely reported in the *Northern Territory News*, and the ship and her crew were feted.

The Master, Peter Maw, was particularly well treated, and recalls two happenings that made this portcall memorable. The first relates to the publicity mentioned in the previous paragraph. A reporter was particularly impressed by GLORIA on the afterdeck, and was endeavouring to take photographs. Peter pointed out that he would get a good overview of the afterdeck and GLORIA from some way up the mast, and the photographer duly obliged. He then asked Peter Maw to come on to the quay to be included in a picture. After moving around seeking the best angles, the photographer eventually asked Peter to sit on a bollard in the foreground of the picture, and then asked him if he would also smoke a pipe! Clearly, the reporter had preconceived ideas of 'ye olde English seadog', and what would make a good picture.

In part because of his involvement with the photographer but also because he had to await the attendance of port officials, the Master knew that he would be unable to visit the town to buy the obligatory mementoes that his young family expected. When this became clear to one of the officials, he offered to drive Peter into town, waited for him, took him home for a meal (apparently a particularly liquid one!), and then returned him to the ship later in the evening. Not unnaturally, Peter Maw has fond memories of the ship's short stay in Port Darwin.

Geophysical studies continued with another GLORIA cruise, this time again led by Lindsay Parson from IOSDL. The remit of his cruise was:

> ..to undertake a geophysical survey, utilising the GLORIA long range sidescan sonar, and geological sampling, utilising dredging and coring, in the Lau Basin to assess processes of back-arc evolution.

Back-arc spreading is what happens at the edges of the tectonic plates when they are moving apart, rather than together as in the previously reported cruise. The plate separation results in molten material from the earth's core welling up and cooling to create new ocean floor. Although not a topic studied by scientists on RRS *Charles Darwin*, where this upwelling takes place can result in areas of exceedingly hot water which support unusual life forms, such as 8-15 foot worms, which are not only of interest to marine biologists but also provide material for televisual nature programmes and writers (*see* Victoria Kaharl, cited on p.40, for a longer description).

An area of back-arc spreading, like the Lau Basin, is of interest to more than just the geophysicist or geologist. Because the seafloor is being changed by the material upwelling from the magma, the topography of the seabed may also be changing significantly. Such changes are likely to be on decadal or centennial timescales, rather than annual, but even so it is essential for accuracy of mapping such changes that the current record represent the actuality as closely as possible. Parson *et al* (1992, 1994) deduced from their GLORIA data that the ocean ridge known as the Central Lau Spreading Centre is propagating southwards at 120 mm per year - ie it is as if some unseen giant hand were tearing open the seabed from north to south at a rate of about 10 mm (or just less than 0.5 in) each *month*. Magmatic material rushes to fill this tear, but such are the pressures and temperature stresses that arise when this super-hot material meets the low temperatures of sea-bottom water that the seabed locally may be deformed. Thus the GLORIA data were used, in conjunction with data from several other sources, to produce a new bathymetric map of the area (Parson, Hawkins and Hunter, 1992).

The bathymetry shows depths between 2,000 and 3,000 metres, and it is from these depths that the scientist wishes to wrest the secrets of the geophysical evolution of the area. GLORIA offers the equivalent of an aerial picture, using reflected sound rather than light, but the scientist would still like to know whether his interpretation of that picture is realistic or not. To do this, he requires material from both the seabed and below it, to confirm his hypothesis, and he uses the coring and dredging techniques referred to in Chapter 3. The core samples taken on this cruise confirmed the scientists' deductions from the GLORIA records, adding to the impression of a successful cruise.

But Lindsay Parson recalls two very special people from this cruise. They were the two observers. It is a requirement on every cruise where research is to be carried out in the territorial waters of a nation or state that at least one observer from the state is aboard for the cruise, together with other conditions about access to data that need not bother us here. Since CD33/88 was to carry out surveys in the Lau Basin, broadly between Tonga and Fiji, an observer from each state was on board. Lindsay describes them - 'the greatest pair of observers I have ever sailed with...Siamone Helu from Tonga, a Sherman Tank of a man, who never saw the need for using the Hiab (crane) on the afterdeck when hauling in the full dredge bags - why not just pick it up and drag it over by hand. But a gentle giant, in every sense of the word, and a great pal to have a run ashore with ... you knew you would never be in peril. And Peter Rodda, our guest from Fiji, a wonderful classical land geologist of many years, who could have cleaned up on Mastermind specialising in Fijian natural sciences. But as I recall, with a pathological hatred of popular music (ie, where 'popular' means any). I still smile to myself as I remember watching Peter tearing up Kleenex to stuff in his ears in order that he could resolutely endure the four hour aural purgatory of an echo-sounder watch'.

How far Lindsay's comments about Peter Rodda's musical appreciation (or lack of it) are justified can be questioned with his (Parson's) final recollection of the cruise. On his office wall hangs a finely crafted and mounted 'gold disc' given to him by all of the ship's engineers. This was 'in memory of my encouragement, aiding and abetting of the Darwin 'rock and pop' fest in the lab during the cruise'. It may be that Peter Rodda's ears were habitually tuned to either the quiet of Fiji or more classical music, with a more specific definition of 'popular' that he abhorred.

This cruise ended in Auckland, New Zealand, having sailed through the Bay of Islands where once again its track crossed that of its namesake. Here it was that Charles Darwin's thoughts were already turning towards home, for in a letter to his sister Caroline, he writes:

> You will see that we have passed the Meridian of the Antipodes & are now on the right side of the world. For the last year I have been wishing to return, & have uttered my wishes in no gentle murmurs; but now I feel inclined to keep up one steady deep growl from morning to night. .... there is no more Geology but plenty of seasickness...

It was not seasickness but concern about the possible harmful effects of imported food and food waste that led the officials in Auckland to impose strict quarantine conditions on ships arriving in the port. Of these, the most obvious was that the ship's refrigerators were sealed locked, and the Master ordered not to open them upon pain of a swingeing fine that would be imposed on him personally. All fresh meat etc. was to be provided from shore supplies. All went well, until one night a member of the ship's catering staff arrived back on board after a 'bender' ashore, felt hungry, and raided the refrigerator for a steak. It was then that one of those co-incidences arose that would not be believed in a work of fiction. When Peter Maw, the Master, had gone ashore soon after arrival in Auckland he literally bumped into a friend of many years standing with whom he had lost touch. It transpired that this gentleman worked for the New Zealand Department of Agriculture, and it was he who detected the transgression of the food regulations noted above when on an inspection visit. Fortunately, Peter was able to point out to him that only a single steak had been removed from the refrigerator, that the residue was still on the plate in the galley, so that no harm had been done, and a fine would be a bit extreme in such circumstances. Common sense and friendship prevailed over too-rigid an application of the rules, and Peter's bank balance was not thereby depleted because of the cavalier attitude of a crew member in his cups.

This part of the western Pacific Ocean is of particular interest to physical oceanographers because it is thought that it may hold the clue to some of the mysteries of El Niño, the Christ Child. This curiously named phenomenon is so called because it is more often than not experienced around Christmas time off the coasts of western South America. El Niño is the term used to describe the occasional strong warming events which involve the entire equatorial Pacific and which prevent nutrients from the deeper cold water from maintaining the anchovy stocks on which the local South American fisheries industry depends. They occur approximately every four years, but there have been significant occurrences seven times in the past 100 years, with one of the most severe occurring in 1982-83. In that year, El Niño caused flooding or drought in 12 countries, resulting in thousands of deaths and ruining billions of dollars worth of property. In Southern California, for instance, the annual rainfall tripled and the coast was battered by storm-force winds and destructive 3-5 metre high waves. A similar weather pattern was experienced at the start of 1995.

Normally, the eastern equatorial Pacific is a region of high atmospheric pressure, while the western equatorial Pacific is one of low pressure. The wind flow from east to west 'drags' the warm water away from the western coast of South America, and this is replaced by colder water welling up. This colder water is rich in nutrients, and makes the coastal waters of Peru and Ecuador some of the most productive in the world.

Two cruises were planned to the western Equatorial Pacific to investigate the factors that might contribute to the generation of El Niño events. Their aim was to improve understanding of the interactions between the atmosphere and the upper ocean in this climatically sensitive region of the world's ocean, with the aspiration to improve predictive models for El Niño.

The specific aim of the first cruise (CD32/88) was formally:

> to undertake studies of the heat and momentum budgets of the mixed layer in the western equatorial Pacific Ocean of the upper 300 metres in the vicinity of the Equator at 165°E.

whilst the later cruise (CD34A/88) aimed:

> to study, in the year following an El Niño event, the dynamics of the near-surface currents of the western Pacific and the structure of the salinity stratified near-surface layer.

Both cruises used similar techniques and equipment, although they covered slightly different areas of the ocean. The first began with SeaSoar and ADCP surveys of a section of the ocean from 155°E to 165°E (approximately 1,000 km) along the Equator, and another which incorporated CTD measurements from 3°N to 3°S (600 km) at 165°E. CD34A/88 surveyed a track covering an area bounded by 7°N 125°E, 7°N 170°E, 10°S 170°E, and 10°S 125°E. The investigations were centred on the fundamental characteristics of the ocean - the spatial variation of salinity and temperature within a prescribed volume of water.

An additional series of intensive measurements during CD32/88 were centred upon a relatively small volume of water - approximately

60 km x 40 km x 0.4 km (or roughly 1,000 cubic kilometres). Into this 1,000 cubic kilometres of ocean were thrown (with appropriate care, and on appropriate cables, of course) all the instrumentation available. A string of current meters was deployed below a spar buoy, coupled with which were a thermistor chain and the IOSDL BERTHA system. The ADCP was activated, and SeaSoar was deployed. The results of all these endeavours produced figures of the variability of the salinity with depth, and the presence of hitherto unsuspectedly extensive sub-surface currents.

These data led the Principal Scientist (Richards, 1991) to conclude 'The observations call into question the ability of present numerical models ... to adequately represent the response of the upper ocean to changes in atmospheric conditions - a necessary requirement for the successful prediction of El Niño events'.

But this cruise was not all hard work: it also produced perhaps more than its fair share of noteworthy personal stories. Of the greatest long-term significance was that a nascent romance blossomed in the sun and heat of the equatorial Pacific, and two of the scientific team - one man and one woman, it should be noted in today's permissive climate - were married shortly afterwards.

Not all intentions were, however, so honourable. The bridge-top on RRS *Charles Darwin* (otherwise known as the Monkey Island) is not visible from the decks, and because it offers a relatively large clear horizontal space in the sun it is a favourite location for sunbathing. Working on and around the Equator, the opportunities for sunbathing were legion, and scientists not on watch made use of these. What they overlooked, particularly during the earlier part of the cruise, was that the scientific remit also included a sky observation camera with a 180° lens. It is not known whether the resulting negatives and prints achieved a cult, voyeur or monetary value, but certainly there were some red faces on the cruise not caused by sunburn!

It was also not revealed whether it was the owners of these red faces who tied the Principal Scientist to the afterdeck and dowsed him in seawater, but unusually this appears to have been the only ceremony on board, in spite of the fact that most of the scientists were 'Crossing the Line' for the first time. Perhaps the Principal Scientist made a mistake in granting his team a day's rest in the middle of the cruise to allow them to unwind.

Finally, Cruise CD32/88 ended in Suva in Fiji. At that time, April/May 1988, Fiji was in a state of political flux. The Governor invited some of the scientists and crew to a 'sundowner' at his residence. Whilst enjoying the traditional drink and literally watching the sun sink below the horizon, the party were perturbed to see a fully uniformed sergeant of the Fiji services, armed with a sub-machine gun, come through the gates of the residence and stride towards them. Just before reaching the party, the sergeant turned sharply left and walked around the side of the house, whence he disappeared for some 30-45 minutes before re-appearing and walking out through the gates. Subsequent enquiries established that his purpose was no more nefarious than visiting his girl-friend, who was the Governor's cook. Political flux or no, the Fijians encountered by the scientists and crew from RRS *Charles Darwin* showed notable friendliness. So much so that Peter Maw recounts that, although the ship was berthed only a few hundred yards from the hotel being used by those arriving and departing the ship, it could take up to three-quarters of an hour to walk this short distance because every local wanted to stop and chat.

Peter also tells the story of his departure from Fiji with some amusement, an emotion which has only developed with hindsight. Peter admits to being terrified of flying, a fear which even his many uneventful flights have been unable to assuage. On this occasion he had to take an internal flight to reach the international airport, and the local plane was a relatively small one with an unpressurised cabin. As it came in to land, the cabin filled with what Peter took to be smoke, and he was filled with alarm. As none of the other passengers nor the stewardess seemed to be taking any notice, he sat with teeth clenched and hands gripping the arms of his seat, convinced that he alone knew that his end had come. When the plane did land without incident, he asked what the problem had been, and was told 'No problem, sir. We normally get condensation in the cabin on this flight'. Peter remains convinced that if God had meant man to fly, He would have given him wings.

The second cruise of these two became a high profile one, although the Principal Scientist is at a loss to understand why. As noted above, the techniques used were identical in both, but Cruise 34A generated a telex to *The Times*, in which the principal comments were:

The justification for the work stems from the efforts of meteorologists and oceanographers to understand and predict the occurrence of El Niño events. These events, the product of a strong coupling between the ocean and the atmosphere, lead every few years to drought in South East Asia and floods in Central and Southern America. As well as having serious consequences for the populations involved, the El Niño events are proving to be excellent testbeds for developing ideas concerning the effect of the ocean on climate.

Present evidence indicates that El Niño events are triggered in the equatorial ocean north of New Guinea, but the details of the triggering mechanism and especially why the events do not occur more often are not understood.

Early results indicate that a key feature of the region are the sharp fronts that develop between the fresh North Pacific surface waters being advected eastward by the North Equatorial Counter Current and the more saline and dense waters of the westward-flowing Equatorial Current. The equatorial waters sink below the fresh North Pacific waters releasing energy and, because of the earth's rotation, appear to generate eddies with a diameter of a few hundred kilometres (in a similar manner, a series of depressions can form along large scale atmospheric fronts). If such an eddy became quasi-permanent then it might result in a pool of even warmer sea surface temperatures which, by causing increased atmospheric convection, could then trigger an El Niño.

The present survey took place in September, a favoured time for generating El Niño events, but in a year following an El Niño when the probability of triggering a new event appears to be lowest. The full potential of the work thus will not be realised until a comparison can be made with similar surveys carried out in years in which El Niño's are triggered.

Yet again, this report points up the two facts of significance highlighted earlier - firstly that the scientists used the availability of the ship in a particular area to add to limited data available from that area, and secondly that before scientific results, however good, can be used predictively they have to be set into context by comparison with a wider data set.

This cruise ended in Papeete, on Tahiti on 11 October 1988, but the ship stayed only two days to change scientific teams. Almost 153 years earlier, HMS *Beagle* had arrived from the other direction for an eleven day stay, and Charles Darwin, after noting that Tahiti was '..an island which must for ever remain as classical to the Voyager in the South Sea..' spends some pages of his diary recording his impressions of both the people and the botany of the island. His colourful language '... It has been remarked that but little habit makes a darker tint of the skin more pleasing & natural to the eye of an Europaean (*sic*) than his own color. To see a white man bathing along side a Tahitian, was like comparing a plant bleached by the gardener's art to the same growing in the open fields...' might excite comment in today's ethos of 'political correctness', but should be seen as emphasising the differences in our own attitudes to both subjective and objective recording.

9
Tonga

Almost by definition, the effective performance of oceanographic research in foreign waters requires the co-operation of the nations or states in whose waters such research is to be carried out. Even where the research itself is in international waters, it is probable that the ship will require to use a foreign port, again requiring permission. The mechanisms for obtaining these permissions are already in place, with the UK being represented overseas by members or appointees of the Foreign and Commonwealth Office (FCO). Over the years, RVS has built up a high degree of rapport with relevant staff at the FCO, and regular liaison meetings between the two organisations ensure that potential problems are recognised well in advance, and that each is fully aware of the other's needs.

It was in the normal course of such liaison that the FCO became aware both of the planned circumnavigation of RRS *Charles Darwin*, and in time of her specific itinerary. When it became apparent that this would take the ship into the Pacific Ocean during 1988, the FCO asked RVS if RRS *Charles Darwin* could represent the UK at the 70th birthday celebrations of His Majesty King Taufa'ahau Tupou IV, King of Tonga. The Island Kingdom of Tonga has had long and friendly ties with the United Kingdom - many older people will remember the delight shown by the King's mother, Queen Salote, when driving in an open carriage in the rain that attended the coronation of Queen Elizabeth II in 1953 - and because of its principally Christian ethic, the King's 70th birthday was imbued with the Biblical meaning attaching to 'three score years and ten'. There was therefore special significance attaching to a UK presence at the celebrations. After due deliberation, it was decided that such a visit could be made without major disruption to the scientific programme, and it was accordingly scheduled.

This was an unique event, and back at Barry there was much scratching of heads about what should or should not be done to make the visit a success. One of the things that was done was to invite to Barry a senior member of the Tongan Police Force, who was studying law at the University of Wales College of Cardiff, to seek firsthand information on the King's likes and dislikes. It emerged that the King was very fond of firework displays, and so an order was sent to the Master of RRS *Charles Darwin* to purchase suitable quantities of fireworks. As the Master subsequently related, trying to buy significant quantities of fireworks in Auckland in June was not the easiest of tasks that had been handed him! But he achieved it, as will be seen below.

Meanwhile, another more formal question had to be answered - what birthday gift could the ship give to the King? The official gift from the UK to the people of Tonga to mark the occasion was an oxygen system for the Island's hospital, but NERC was not expected to match that. So what could it be? Again, discreet enquiries revealed that the King had more than a passing interest in matters relating to the evolution of the Tongan Islands, and so what could be more relevant than a GLORIA sonograph of the area? This was duly prepared, and by one of those unforeseen co-incidences, involved a family link. One of the NERC geophysicists involved in collecting and assembling the GLORIA data was one Quentin Huggett, from the Institute of Oceanographic Sciences Deacon Laboratory. In Auckland, it was necessary to mount and frame the GLORIA image, and it just happened that the company chosen to do this was run by his brother, Steven Huggett! When this picture was presented to His Majesty, he walked to one side of the reception room and compared the topography revealed by GLORIA with a representation of the seabed that had been given to him some time previously, and is reported to have remarked that the two seemed to agree remarkably well, thus confirming both his interest in the topic and the wisdom of NERC's choice of gift.

RRS *Charles Darwin* arrived in Tonga on 2 July, in preparation for the principal celebrations on 4 July. Also there were two French Naval vessels, and one ship from each of the Australian, New Zealand and US

PROGRAMME FOR THE ROYAL FIREWORKS

---

1. ROCKETS FROM FOR'D, MIDSHIPS & AFT
(Rocket launching stations)

2. SPARKLERS.
Scientific party under direction of
Eric Darlington semaphore:-

H A P P Y   B I R T H D A Y

3. ROCKETS FROM FOR'D, MIDSHIPS & AFT

4. POP UP PYROTECHNICS PLACED ALONG THE LENGTH OF SHIP SET OFF SIMULTANEOUSLY

5. ROCKETS FROM FOR'D, MIDSHIPS, AFT

6. Firework display by HMNZS "TUI"

7. Hari Kari of C.D. Master?

8. Vessel sails in haste for Auckland

navies. The naval ships formed a review line some 2,000 yards offshore, whilst RRS *Charles Darwin* was placed at the head of a line of civil ships closer inshore. No time was wasted getting into the swing of things, and on the evening of 2 July, the Master (Sam Mayl), the Head of RVS (Dr Leonard Skinner), and the UK High Commissioner (Paul Fabian) hosted a reception on board the ship, attended by about 100 guests.

3 July was a Sunday, and was observed in the appropriate manner, with church services having a special character and being more fully attended than normal. Then on 4 July the partying really began! The day started with a review by HM the King of the civil and naval ships, followed by various marine activities. From midday, the feasting began, including the traditional suckling pig and other exotic barbecues. These feasts were spread across the island, and were laid out under awnings of palm leaves, with all the guests - both visitors and the local populace - seated on the ground adjacent.

In the evening, the ships, which were anchored offshore close to the Royal Palace, were expected to contribute to the celebrations. RRS *Charles Darwin* led off, and the Master's instructions convey the sense of occasion *(see panel opposite)*:

The Master should have had no cause for concern. The reports reaching RVS indicated that RRS *Charles Darwin* had acquitted herself well - as the High Commissioner reported '..Her firework display on the 4th considerably outshone those of her naval cousins'. Of course, the ship's strategic position at the head of the line, offshore from the royal palace, and between the palace and the line of warships, helped. The fireworks contributed by those warships *appeared* from the shore to have been fired by RRS *Charles Darwin*, and this could well have influenced the formal and informal plaudits that followed!

The visit of RRS *Charles Darwin* also marked the centenary of the visit to the Friendly Islands of HMS *Egeria*, and was also the first visit by a UK Royal Research Ship to Tonga since that of HMS *Challenger* in the early 1870s.

The (then) Head of RVS (Dr L M Skinner) gave an extended interview to the Anniversary Issue of *Tonga Today*, in which he and some of the scientists on board the ship gave a publicist's view of what marine science is all about. An extended quotation gives a 'flavour' of what was said.

...our lives depend on (the ocean) in ways we are not even aware of. For instance, the ocean is the chief driving force of the weather, acting almost as an incubator around the land, keeping the air cold when the currents are cold and the air warm when the ocean is warm. And when you disturb the balance of the ocean, you disturb the land we live on too. An example of this is the sinking of Venice and Bangkok caused by the use of ground water which helped to keep those areas buoyant and thus afloat.

So what have the scientists learned about Tonga from their research of the ocean in the Pacific region? ... Tonga is not one of those islands that is sinking, but is actually rising. And Tonga is moving. This is due to what scientists term a mid-ocean ridge between Tonga and the coast of South America. As explained, the earth is very hot, and heat rises. Thus, somewhere the earth's surface must give. In many places this happens beneath the ocean, as ridges of earth and rock appear. Because the material that is displaced has to go somewhere, it moves (in this case west) and pushes everything in its path along with it.

It is this type of geological happening which causes whole new land formations, the Darwin scientists explained. For example, India slowly floated until it crashed into Asia, creating the Himalayas. And while these ridges are found in all the oceans of the world and the surface land is thus constantly in motion, the movement in the Pacific is extremely rapid, by geological standards, making the region one of the most exciting places for scientists to work...

(the GLORIA sonograph presented to His Majesty)..shows an area of 400 square kilometres, some 2,500 metres below the surface. What is seen in the picture is the great number of volcanoes on the seabed, volcanoes which will in hundreds of years become new islands.

This chapter has been concerned with one aspect - the 'public relations' - of the role of RRS *Charles Darwin*, an aspect that was and still is important to the continuation of marine and environmental science in a world that seems to place increasing significance on an immediate return from research. By definition, environmental studies are rarely in this category - in most of the relevant disciplines an historical record is essential to gauge whether perceived changes can be attributed to natural or man-made factors. Nowhere is this more pertinent than in matters marine.

Almost all of this painstaking historical data will have been assembled by scientists of all the generations from (or even before) the global efforts of Mr Charles Darwin, but ever since his seminal work *On the Origin of Species* was published, the non-specialist has become increasingly aware of the impact of scientific findings on his or her everyday life. With this awareness has come a wish to understand more of the detail (without necessarily having to take in all the theory!), and it is the interest of national leaders such as His Majesty the King of Tonga which can further stimulate this interest. Whilst the 'public relations' efforts of major (and costly) facilities, like RRS *Charles Darwin*, will be scrutinised closely by those with charge of the purse strings of modern science, and criticised by the scientists themselves as demanding money that could otherwise have gone to support their science, it is almost an essential pre-requisite in a world where science and politics are so closely interwoven. That RRS *Charles Darwin* was able to 'show the flag' was a sign of the times, and is reflected in the formal report of the High Commissioner, when he says:

> I venture to hope that it will not be another 100 years before (RRS) *Charles Darwin* or one of her sister ships returns....I believe much quiet prestige accrued to Britain from the scientific as well as the representational side, with the glamorous Gloria (*sic*) sonar equipment to the fore.

10
Eastern Pacific

RRS *Charles Darwin* was, however, to support 'more geology' on her departure from Auckland towards Tonga, and on that geophysics cruise, Martin Sinha, from the University of Cambridge, complemented the systems noted previously with an

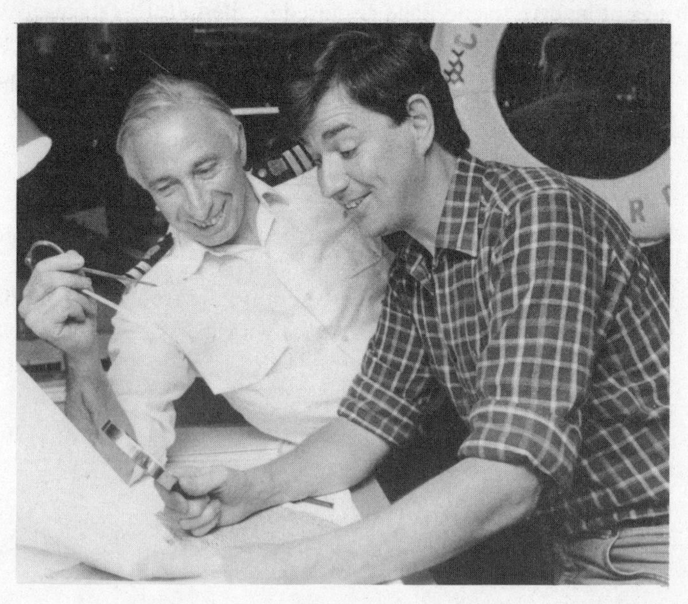

Plate 11: Sam Mayl and Martin Sinha

entirely different technique. In Auckland, prior to departure for Tonga, RRS *Charles Darwin* attracted the, by now usual, press interest, and Martin and the ship's Master, Sam Mayl, were photographed preparing their strategy for Cruise CD34/88 (Black and White Plate 11).

The *New Zealand Herald* noted:

> (RRS *Charles Darwin*) will survey a 200km submarine volcanic ridge between Fiji and Tonga. Once there, the head of the 17 member scientific team on board, Dr Martin Sinha, hopes to confirm the existence of a gigantic lake of molten lava estimated to be about 3 km beneath the ocean floor. If the team is successful in finding the undersea lake, it will have proved correct the educated guesses of scientists who have predicted its existence for over 20 years.

To set the scene, Martin Sinha himself writes:

> The Lau Basin is a young (less than 4 Ma old) back-arc basin of roughly triangular shape, bounded to the west by the Lau Ridge, and extinct, volcanic, island arc, and to the east by the Tonga

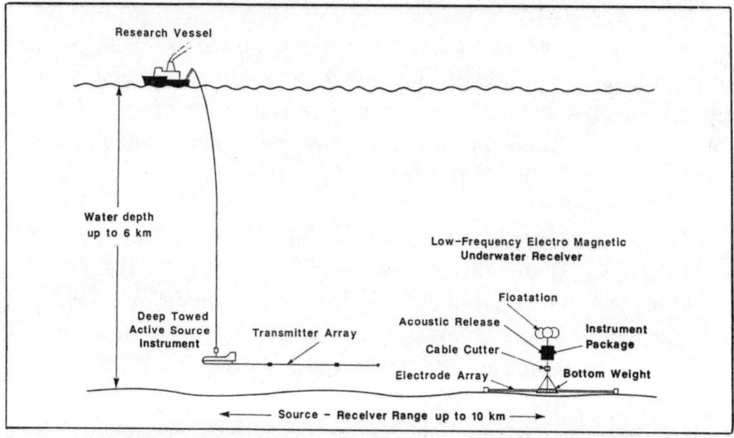

Plate 12: Schematic arrangement of an active source electromagnetic sounding experiment, showing the deep-towed active source instrument, the surface research vessel and a low-frequency electromagnetic underwater recorder (not to scale).

> Islands - a currently active arc system... Subduction of the Pacific plate is occurring beneath and to the east of the Tonga Islands, at the Tonga Trench. The Lau Basin is therefore underlain at depth by the westerly-dipping, downgoing, subducted Pacific slab.

The Valu Fa Ridge is a segment of a spreading centre - ie that area of tectonic plate evolution where the plates are moving apart, and new material is being created by magmatic upwelling. The ridge has been developing for less than 1 million years, and has a spreading rate of between 60 and 70 mm per year, and Sinha's specific interest is in magmatic upwelling, and the possibility of quantifying the extent of chambers of magmatic material immediately below the seabed. He reasoned these could be detected by their electro-magnetic signatures, and accordingly he developed an instrument for the measurement of such signatures. Sadly, on this cruise Sinha (1989) was to report.. 'The electro-magnetic sounding work was unsuccessful. A series of equipment failures resulted in only a small amount of data in a severely limited frequency range being recorded. The data obtained will be insufficient to allow an analysis to be made of crustal conductivity structure. Two seabottom EM recorders were lost'.

This extract from the Cruise Report highlights the fact that the course of science does not always run smoothly, and that a combination of equipment problems and the inherent resistance of the ocean to the probing of its depths can upset the best laid plans, a theme that is continued in the following extracts...

> On completion of the gravity profile, began deployment of DASI[22] ... At 206/0652[23] overheating elements in the DASI power supply caused smoke from scorching paintwork, which triggered the smoke alarm system. The elements were moved to a different location, and a larger fan was installed for cooling them.
>
> DASI and streamer brought back to the surface. We found that, during the deployment, the streamer had become wrapped around the deep tow cable causing serious damage to one of the electrode cables and one of the flotation sections....

Test and calibration transmissions were interrupted by a major failure....The period was enlivened by a flood of seawater into the Controlled Temperature Laboratory, which we were using as the .. electronics laboratory. The flood was caused by the ship's engineers turning on the non-toxic seawater supply pump. A tap connected to this supply over the sink in the Controlled Temperature Laboratory had been left on, and since the sink had been covered by a wooden bench top for the duration of the cruise, the resulting jet of seawater could not escape down the plug hole. Worse, the bench top also prevented the tap from being turned off. ... Fortunately, no serious damage was done - mainly because the electronics packages, which would have been directly in the path of the water, were sitting on the seabed at the time.

DASI grounded on the seabed on the side of a seamount, during a turn at the northern end of the planned deep tow track...The deep tow itself and conducting swivel had received only superficial damage. However, the outboard 650m of the deep tow cable had suffered significant damage to its armour, and the deep tow antenna was gashed and deeply abraded in numerous places. It was clear that, in the time available, no further deep tow operations would be possible.

However, to offset all this apparent doom and gloom, Sinha notes

The seismic experiments were extremely successful, producing 1,300 km of digitally recorded, 4-channel reflection profiles...The seismic streamer was the object of a great deal of painstaking preparation by RVS staff at the start of the cruise. This effort paid off in both the complete reliability of the streamer once the profiling had started, and the high quality of the data that we collected.

This led Collier and Sinha (1992) to deduce that the Central Valu Fa Ridge was not a single segment of a back-arc spreading centre, but could be categorised into three 8-12 km sections, each with its own distinctive geophysical characteristics.

Roger Searle, then at IOSDL, but currently Professor of Geophysics at the University of Durham, continued the geophysical theme with a cruise which had as its formal remit:

> ...to undertake studies of the present configuration and past tectonic history of the Easter and Juan Fernandez microplates on the East Pacific Rise, by mapping plate boundaries and inferring spreading rates and directions. A brief reconnaissance of the Chile Fracture Zone and Chile Ridge will be conducted on the final passage to Valparaiso.

The Master's report on the cruise gives a terse summary of what science means to the non-scientist - 'Equipment used included GLORIA, 3.5 kHz fish[24], gravity and magnetics. This was deployed off Papeete and recovered 6,445 miles later off Easter Island'. But 35 days of science should not be so lightly dismissed.

Plate 13: Sidescan sonar mosaic of Easter microplate. Most the of the area is covered by GLORIA data (40 km-wide swaths).SeaMARC II sidescan data (10 km-wide swaths) are used to provide complete coverage of the area around Pito Deep (north-east corner of microplate boundary) and the southern end of the south-west rift, and to partially fill other areas of incomplete GLORIA coverage. Darker tones show stronger acoustic back-scattering, which usually indicates less sediment cover and therefore younger sea floor. Data have been digitally processed to correct for geometric distortions, to equalize the effective gain at different ranges, and to remove line dropouts.

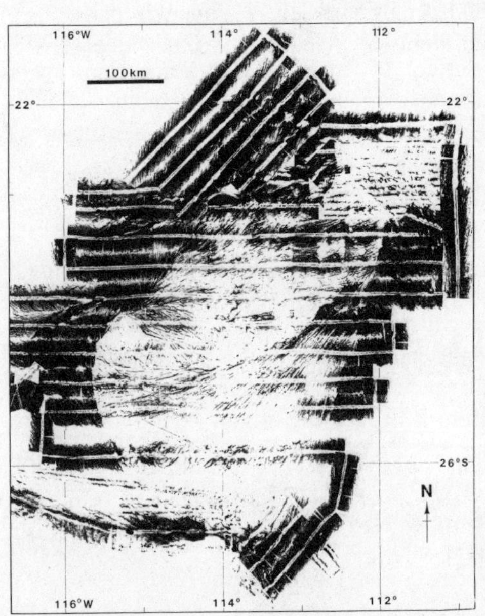

In earlier chapters, we have referred to tectonic plates, clearly implying that there are relatively few such structures that make up the earth's surface. Reference to the map at Colour Plate 2 will show that this is superficially the case. However, earlier chapters have also shown that much of the work of the geophysicists and geologists is concerned with inhomogeneities in the plates, and the present cruise continues this theme. Whereas the major plates have principal dimensions measured in thousands of kilometres, the 'microplates' to be studied on this occasion are both less than 500 km approximate diameter.

GLORIA provided a technique that would allow an entire microplate to be surveyed on one cruise, and was fruitful in defining fully and accurately, for the first time, the microplate boundaries. This was because GLORIA, when operated in good conditions, can produce a sonograph (ie, an underwater equivalent of an aerial photograph) some 30 km either side of the ship's track, and with uninterrupted steaming, as on this cruise, can survey an area the size of Wales every day. Searle *et al* (1989) state, 'One of our most important findings is that all of the microplate boundaries are evolving rapidly.'

The Easter microplate lies between the Pacific and Nazca plates, and the Juan Fernandez microplate sits at the triple junction between the Pacific, Nazca and Antarctic plates. As we have already seen, the tectonic plates that comprise the surface of the globe (the lithosphere) are in constant motion, with an average speed relative to one another of about 30 kilometres per million years. On the East Pacific Rise, where these two microplates are situated, this relative motion between the plates is very much faster, and is thought to have worked on a small piece of broken plate, and rotated it like, as one early model proposed, a roller-bearing between two surfaces. In fact, the microplate is not a perfect circle, and material is being added to and lost from it all the time, so that the model is more analogous to an irregular cam being moved by shear forces.

Searle *et al* (1993) used the data gathered on this cruise, amalgamated with complementary data gathered by international colleagues, to develop this model and from it deduce the origins of the microplates. They showed that the roller-bearing model correctly predicts the current rate of rotation of the Easter microplate. When applied to the Juan Fernandez microplate, the data shows that that microplate was indeed rotating at the predicted rate until about 2 million years ago, when it moved out of the sole influence of the Pacific-Nazca plate forces.

The results from this cruise also attracted wide publicity, and an article in *The Independent* on 9 January 1989 was headlined '*Charles Darwin's new voyage in the Pacific*' with a subtitle '*Roger Searle on an expedition which has shed light on the sea's evolution'*. After outlining the basics of tectonic plate theory, and the efforts of the cruise in mapping the structures of the microplate, Searle outlines what he believes will follow. This was:

> ...we now believe that we will be able to reconstruct the microplate's entire history. If so, for the first time researchers will have a complete record of the history - and possibly origin - of a tectonic plate, albeit a small one. This will be of significance to geologists for a number of reasons. First, it will provide a test of the usefulness of plate tectonics at a very small scale, and in doing so may strengthen that theory as a means of predicting geology elsewhere in the world, including such economic aspects as the location of mineral deposits.
> Another important outcome should be a better understanding of the way in which plate boundaries are re-organised world-wide. In the past the boundaries between major plates have changed radically. It is beginning to appear that many of these changes involved one or more temporary microplates. However, the record of these older microplates is sparse and difficult to piece together, so a complete description of a modern 'working' microplate will be a valuable aid.
>
> The work may also provide a window into the internal workings of the earth. Although it is believed that plate motions are associated with slow 'convection' of the rocks deep in the earth's interior, these motions have been hard to discover because the plates behave like a mechanical blanket, hiding them from view. Because of the high spreading rates around the Easter microplate, the lithosphere is particularly thin and weak there, and should reflect the deep convective movements more faithfully than elsewhere.
>
> Recently, geologists have come to think that the pattern of segmentation into which spreading centres are broken up by offsets such as transform faults and propagating rifts may actually

reflect the segmentation of these deep currents. If so, the history of the microplate should tell us more about the way these movements are organised, how they change, and ultimately perhaps how they control sources of minerals and hydrothermal energy.

It is intriguing to contemplate the marvellous complexity of this small piece of earth, and its rapid evolution to its present structure in about the time that man has inhabited the world.

Roger Searle has revisited this press article, and has suggested that several of the ambitions posited have been, or are on the way to being, realised. Point by point, he offered the following:

.. we now believe that we will be able to reconstruct the microplate's entire history.

Ruth Rusby, a doctoral student on CD35/88, carried out the major analysis of the CD35/88 data for her PhD thesis (Rusby, 1992), the bulk of which is about to be published as Rusby & Searle (1995). She determined the detailed history of the microplate, and suggested that it formed about 5.25 million years ago as a small 'chip' broken off the major Nazca plate.

.. will provide a test of plate tectonics at a very small scale

Rusby found that most of the microplate's motion could be adequately described by rigid-plate tectonics, but that the northern boundary included a wide zone of seafloor that has been deformed by compression and shearing (Rusby & Searle, 1993).

.. a better understanding of the ways in which plate tectonics are organised

One outcome of CD35/88 was a poster presentation of the early results to the 1988 Fall Meeting of the American Geophysical Union, just a few months after the end of the cruise. At that Meeting, some American colleagues of Searle's suggested some follow-up work, as a result of which R/V *Maurice Ewing* supported a comparable study of the Juan

Fernandes microplate in 1991. The structure and behaviour of the two microplates are remarkably similar (Searle *et al*, 1993), which gives Roger some confidence that his aim to have a 'complete description of a modern working microplate' has been achieved.

..may also provide a window into the internal workings of the earth

With predictable academic caution, Roger suggests that 'it is early days to be positive that this ambition will be fulfilled.' However, he notes that the fact that the behaviour of the very weak lithosphere around Easter can be so well modelled by the plate theory suggests that the underlying mantle motions here may be quite simple, unlike the richly-structured, shallow 3D convection pattern that has been suggested for some other parts of the ocean.

...the pattern of segmentation ... may reflect the of segmentation of these deep currents

At the microplate boundaries, the segmentation is at least partly controlled by the tectonic stresses in the plates. The pattern of stresses change as the microplates move, and the microplate boundaries continuously re-organise themselves into spreading segments that are aligned at right angles to the main tensional stresses (Searle *et al*, 1989, 1993).

...the history of the microplate should tell us ... how they control sources of minerals and hydrothermal energy

This question led to Roger participating in another collaborative international cruise, this time on the French vessel N/O *Nadir*, and its submersible, the *Nautile*. The cruise, titled 'PITO', after the Polynesian nickname for Easter Island, 'the navel of the world', was led by M. Jean Francheteau. The two scientists discovered several new hydrothermal venting sites around the microplate, and the results are, at the time of writing, being closely studied.

In summing up the results of RRS *Charles Darwin* Cruise CD35/88, Roger Searle writes:

As well as the detailed study of the microplate, the GLORIA and other observations made on passage from Tahiti have yielded important new information on the geology of this region. First they enabled us to define the precise position of the Society Islands 'hot-spot' - the region where one of the plumes of hot rock rising from the Earth's mantle intersects the seafloor (Binard *et al*, 1991). Further, they showed that mid-plate volcanism is unusually extensive in this area, and finally that the accepted scenario of seafloor spreading history will need some modification (Searle *et al*, 1995). All in all, RRS *Charles Darwin* Cruise 35 has proved to be extremely productive.

This cruise ended in Valparaiso, Chile. It was here that HMS *Beagle* stayed for some months in 1834, and the naturalist took himself off into the country exploring and researching. However, it is likely that conditions have changed in the intervening years, as Charles Darwin writes

> Valparaiso is a sort of London or Paris, to any place we have been to - it is most disagreeable to be obliged to shave and dress decently. We shall stay here two months, instead of going Northward, during which time the ship will be refitted & all hands refreshed.

Unfortunately, the three and a half month stay in Valparaiso did not provide the refreshment that Charles Darwin needed - he spent almost half the period in bed with an unidentified illness contracted during one of his expeditions into the mountains. When RRS *Charles Darwin* arrived in Valparaiso it was also time for her annual refit, and one or two of the support staff took time for 'refreshment'. One of them recalls visiting the local horse-racing track one evening clutching a fistful of local currency. There were however certain restraints on over-indulgence, and no bet larger than the UK equivalent of 10p was allowed on the course. Whilst this did prevent financial disaster, it also meant that the one member of the party who had a successful evening returned to the hotel with a carrier bag full of money with a conversion value of about £5. Whether Charles Darwin would have approved of either the activity itself or the paucity of the reward we cannot judge.

Meanwhile, his namesake continued to attract attention. At the end of Roger Searle's cruise in Valparaiso, the local paper put a somewhat different interpretation on the origins of the microplate. *Revista del Domingo (The Sunday Review)* notes

> It was in this region (Easter Island), some three million years ago, that a meteorite crashed, leaving a gaping hole some 500 kilometres in diameter. It is thought that the pressure and impact resulted in the formation of a volcano which is still throwing up lava from the bottom of the sea. This upheaval would have changed the climate, exterminating species such as the dinosaur.

Which only goes to show how very careful a scientist must be when explaining his subject to the layman or press. But to be fair, the same article then reports on the next cruise to take place under the leadership of Graham Westbrook, from the University of Birmingham.

> In Chile, the use of this six metre long yellow torpedo (GLORIA) will concentrate on the triple junction area of those sections of the earth's crust facing the Lagos and Aisen regional coastlines.
>
> 'The only point in the world where this collision happens is to the north of the Taitao Peninsula' states Graham Westbrook, leader of the team of scientists on Expedition No 36. 'In the past, the point of impact for the three terrestrial plates lay further south, but it's been proved that with time, and over millions of years, it's been gradually moving northwards'.
>
> The Taitao Peninsula is more raised because it is supported on rocks which could have been thrown up by collision. 'At some time in the future the Peninsula will begin to collapse as a result of the displacement at the point of the triple junction ... but not for several million years yet', he smiles reassuringly!

And so, with that merry quip, Graham Westbrook set sail for the Chile Triple Junction. His objective, formally summarised, was:

..to undertake studies of the effect that subduction of the Chile Ridge, an active spreading centre, has had upon the tectonic structure and sedimentation of the continental margin of Chile, and how the margin has responded as the triple junction formed at the intersection of the ridge and trench has migrated northwards along the margin.

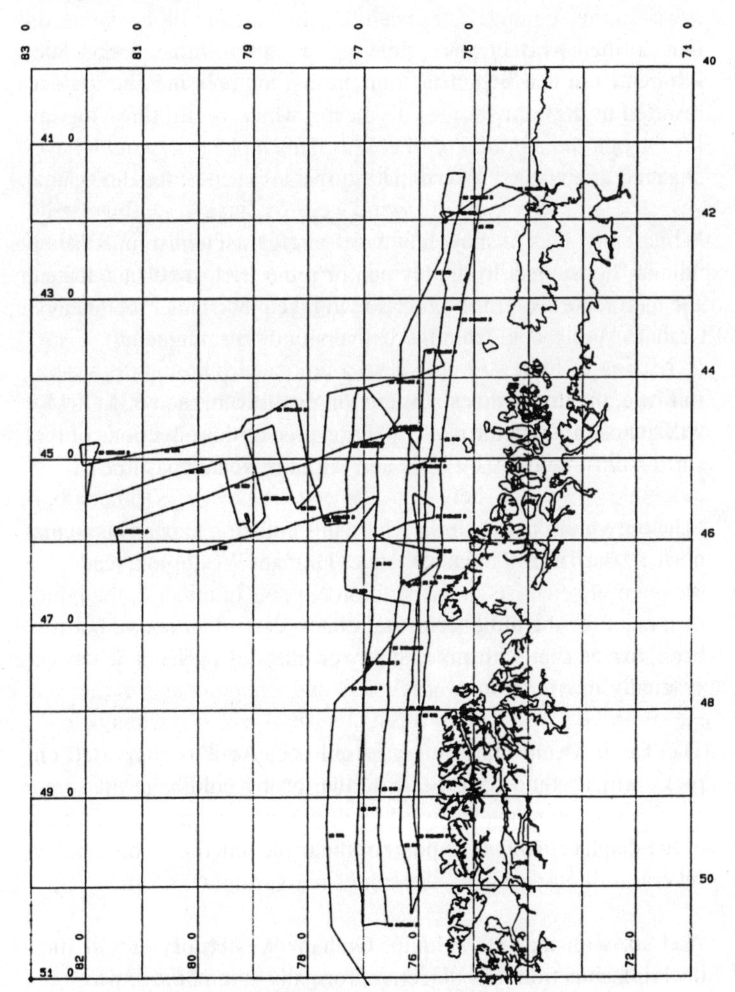

Plate 14: Map of CD36 ex Westbrook's scientific paper

The map below shows the geography and ship's track for this cruise, and sets some of the conditions under which scientists will labour to obtain data to prove, improve or disprove a theory. The working area was off the southwest coast of Chile through December, the so-called 'Austral summer'. However, the comparison with the northern summer is pointed up by the Master's report, which notes:

> On sailing, weather was quite good, apart from a moderate southerly swell which caused the vessel to pitch, making quite a few people feel unwell.
>
> During the evening of 2nd (December), fog set in, causing the vessel to slow down...on 4th, wind westerly force 4, southwesterly moderate to high swell causing vessel to pitch and roll....During 5th, the wind steadily increased from the north until it reached gale force with a rough following sea and moderate-heavy rain...During the night 5th/6th the wind slowly went to NNW increasing to force 8/9 with a very big swell building up, vessel rolling violently at times. During the day of 6th, the wind slowly went to south west, with very little moderation, and a massive swell still running.... (by 0700 on 7th) ...the weather was generally moderate, wind was between NW and SW average force 6 but reaching gale force at times. We had long periods of rain and/or mist, with visibility hovering on poor.

These conditions did little or nothing to deter the intrepid scientists, because the Master's report also noted that GLORIA and the other sampling instruments were deployed and operational for the whole cruise. With GLORIA being towed at several tens of metres below the sea surface, it was unlikely to be disturbed by swell or waves of 'only' a few metres on the surface, and so the team would have made every effort to ensure that the data being transmitted to the ship were properly recorded for later analysis and reporting. Indeed, Westbrook (1989a) reports ..

> The cruise was successful in surveying virtually all of the intended area with GLORIA, although this was at the expense of

the loss of seismic reflection coverage of areas for which seismic reflection was not critical.

However, Westbrook does qualify this comment with:

> This was achieved despite the magnetometer, gravimeter and seismic reflection system developing problems that put them out of action for various periods of time.

But after commenting in detail on the equipment shortcomings, Graham goes on to add:

> The ship performed well throughout the cruise. The ship's officers produced the high standard of competence that one has come to expect of them. The deck crew were distinctly better than on some

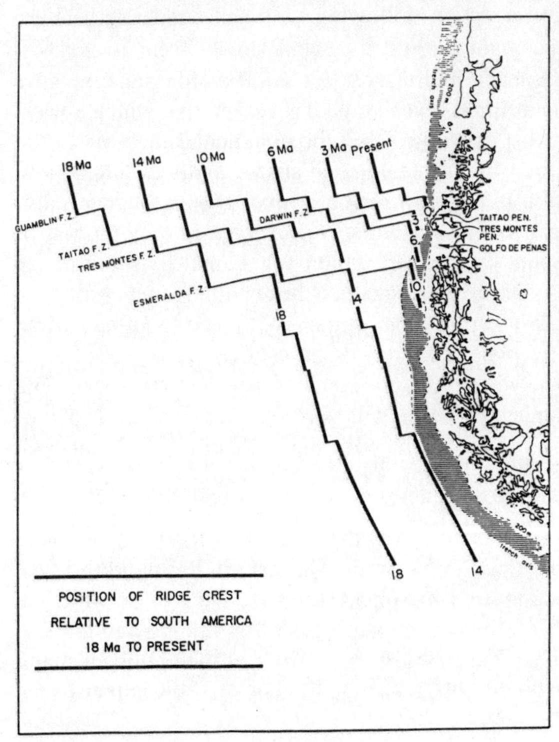

Plate 15: The motion of the Chile Ridge with respect to South America over the past 18 Ma, showing subduction of the Ridge crest between 14 and 10 Ma, at 6 Ma, at 3Ma and at present (from Cande and Leslie, 1986)

cruises in which I have taken part: they were co-operative, cheerful, and did their work well.

In this respect, Professor Westbrook is repeating a philosophy expressed over 150 years ago by the ship's namesake. Keynes noted:

> The achievements of the *Beagle* did not just depend on Fitzroy's skill as a hydrographer, nor on Darwin's skill as a natural scientist, but on the thoroughly effective fashion in which everyone on board pulled together....And anyone who has participated in a scientific expedition will agree that when he wrote from Valparaiso in July 1834 that 'The Captain keeps all smooth by rowing everyone in turn, which of course he has as much right to do as a gamekeeper to shoot partridges on 1 September', he was putting his finger on an important ingredient in the *Beagle's* success.

Scientifically, Graham Westbrook found this one of his more rewarding cruises, because it revealed unexpected results which always excite the scientist. He confirmed that the continental margins to the north and south of the Chile triple junction are markedly different, with the rate of subduction in the north being 4 times that in the south. For comparison, we have commented that the average rate of motion of tectonic plates is about 30 km per million years, and the south of the triple junction the Antarctic Plate subducts beneath the South American Plate at about 20 km per million years. To the north of the triple junction, the Nazca plate subducts at rates that have varied over (geological) time between 80 and 120 km per million years. The GLORIA and seismic reflection data showed that the thickness of sediment in the trench to the north and to the south of the triple junction is about the same, so the volume of sediment being delivered to the continental margin north of the triple junction in a period of time is four times greater than to the south, yet the margin to the north is much narrower and steeper than to the south. The margin to the north is cut through by canyons and has widespread slumping and erosion while the margin to the south is clearly growing outward by folding and faulting the sediment in the trench, piling it up in front of the continent like a bulldozer scraping up the earth. The significance of this is that to the south of the triple junction, nearly all of the sediment on the ocean floor accumulates against the continental margin whereas to the north of the triple junction the sediment is taken deeper into the Earth, down the subduction zone, where it can be added to the base of the South American Continent, and contribute to melting of the Earth's mantle and the formation of igneous rocks that form the volcanic chain of the Andes. Westbrook and Lothian (1994) explore a number of possible hypotheses for the apparent anomalies revealed by this cruise, but conclude that further research is necessary before any of them can be fully developed. They note

> The most obvious and pressing research to be pursued would be to undertake drilling at sites north and south of the triple junction....This would enable the determination of the age of accreted sediment and also the age of the fore-arc basin south of the triple junction, with in the latter case some idea of the history of subsidence and uplift.

This cruise provided information for Leg 141 of the Ocean Drilling Program which drilled five sites on the lower continental margin in the close vicinity of the triple junction in November and December 1991.

At the conclusion of this cruise, RRS *Charles Darwin* headed south towards Cape Horn and Port Stanley in the Falkland Islands, in effect retracing the path followed by HMS *Beagle*. There was, however, one significant change forced by the different political situations of 1833 and 1988. On his arrival in the Falkland Islands on 1 March 1833, Charles Darwin noted his

> ..astonishment, that England had taken possession of the Falkland Islands & that the Flag was now flying. These Islands have been for some time uninhabited until the Buenos Aires Government a few years since claimed them & sent some colonists. Our Government remonstrated against this, & last month the Clio arrived here with orders to take possession of the place... The present inhabitants consist of one Englishman who has resided here for some years & now has charge of the British Flag, 20 Spaniards & three women, two of whom are negresses.

Although, therefore, the politics of the time were volatile, this did not stop HMS *Beagle* from sailing from the Falklands to Valparaiso through the Magellan Channel. In 1988, although some six years after the armed dispute over the Falkland Islands, political susceptibilities were such that a passage through the whole of the Magellan Channel was deemed undesirable, to put no finer point on it. Accordingly, RRS *Charles Darwin* supported multi-channel seismic on passage to Punta Arenas via the western end of the Magellan Channel. In Punta Arenas the scientific party disembarked, and the Chilean pilot took the ship through a number of minor channels to approach Port Stanley in the Falkland Islands from the south west.

The ship's Master, Peter Maw, recalls two events from this trip. The first is that the channels were really minor - at one stage one of the two Chilean Pilots on board (essential because the whole passage had to be under Pilot control) suggested that he, Maw, might like to keep his nerves intact by retiring either to a corner of the bridge or to his cabin as there would be less than six feet either side of the ship. Peter reports that some intrepid crew members actually leaned over the rail and attempted to touch the walls of the channel, and claims that he aged considerably on that trip! The other story is that because these channels were not frequently used, the staff of a Chilean Radio Station sited at the exit to one of them scarcely ever saw another human being. When RRS *Charles Darwin* was abaft this station, one of the Pilots sounded the whistle and such was the interest and excitement of those manning the station that they all ran to the cliff edge waving and cheering, including two individuals who had obviously left their ablutions in a hurry, since they were wearing nothing other than a towel!

And so RRS C*harles Darwin*, sailing the reverse course to its namesake, came to the Falkland Islands and to the harbour at Port Stanley. Obviously much had changed in the intervening 155 years, in terms of both the novelty of the sovereignty claim and the population, but as we shall see both the man and the ship suffered from the constancy of less-than-friendly weather during their stays.

---

[22] DASI - Deep Towed Active Source Instrument - a horizontal electric dipole transmitter, developed by Cambridge University, and towed just above the seabed using the ship's electrically conducting cable.

[23] Time references are given in terms of Julian Day and GMT.

[24] The sensor towed below the ship is a hydronamically shaped body, colloquially referred to as the 'fish'.

11
The Antarctic

RRS *Charles Darwin* was built to a budget which not only dictated her size, but also meant that it was not possible to incorporate ice-strengthening into her construction. Accordingly, when the possibility of undertaking research in the Southern Ocean was being discussed, it could be agreed only under very strict operating conditions which would eliminate, as far as possible, the danger of an unseemly argument between the ship and any wayward ice. The scientists from the British Antarctic Survey accepted this condition, planned their work area accordingly, and the cruise reported here was the result.

The cruise was made up of three legs, each of about one month's duration, this being the practical limit imposed by the ship's design in respect of fuel and food. The waters off Antarctica are inhospitable - always cold, often ice-laden - but like the rest of the world's oceans, cover geological information of increasing interest to scientists. NERC's British Antarctic Survey (BAS) is a multi-disciplinary institute whose research interests encompass all aspects of the Antarctic environment, from the microflora of mosses that survive in the rocky interstices of the icy surface, through the diving habits of penguins and seals, to the geophysical history of the Antarctic landmass and surrounding oceans and lithosphere. From information gathered over the years, including a cruise on RRS *Discovery* in 1987-88, the BAS geophysicists had formed some theories about the geological evolution of the area, but they required the detailed picture of the ocean floor that GLORIA could provide in order to test these theories.

Barker & Pudsey (1991) note:

> The BAS marine geoscience interest in the Scotia Sea region can be divided into three fields:
> (a) regional tectonic evolution;
> (b) subduction-related processes;
> (c) palaeo-oceanography and glacial marine sedimentation.

Thus it came about that Cruise CD37/88 had the following remit:

> ...to continue studies of the regional tectonic evolution, subduction related processes and palaeo-oceanography, and recover and redeploy a series of moored current meters deployed in early 1988

The first leg of the cruise took place in January 1989, and a measure of the problems faced can be garnered from the report of the Master (Peter Maw).

> 03 (January) 2330 GLORIA, magnetometer, etc., all streamed. Weather reasonable, with WNW wind force 5/6, moderate sea and swell....During the night of 6/7 January we encountered our first icebergs...We also had snow showers to go with them...For the next 48 hours we encountered heavy concentrations of icebergs, having at times 50 bergs in the 24 mile range on the radar....During the early hours of 13th, the visibility started to close in. At 0900 the fog was persisting...1330 fog lifted...1500 fog closed down again...1830 slight improvement...by 2300 once more in dense fog...fog continued intermittent until 14 0430, when it finally lifted.

> Over 22 and 23 (January) ..the barometer fell sharply, until at 23 1800 it stood at 946mb and still falling...23 1900..all gear recovered and vessel hove to head to weather in N'ly wind force

7/8...Wind increased to storm force 10 for 24 hours, then slowly started to moderate. By 25 1600 wind force was down to 8/9...By 25 1830 the wind and sea had eased considerably, the magnetometer was streamed and magnetic survey continued.

During the morning of 3 February... the weather rapidly deteriorated until the wind had reached force 9, with a big sea building up... During the day the wind steadily increased, between 1500 and 1700 we had gusts of 70-75 knots, force 12; during the evening it slowly began to abate (until) by 2000 wind was force 8/9...

The main problem we had to contend with on this leg was weather. For the first 10 days the weather was quite reasonable; after that we had a high incidence of fog. When it wasn't foggy, it

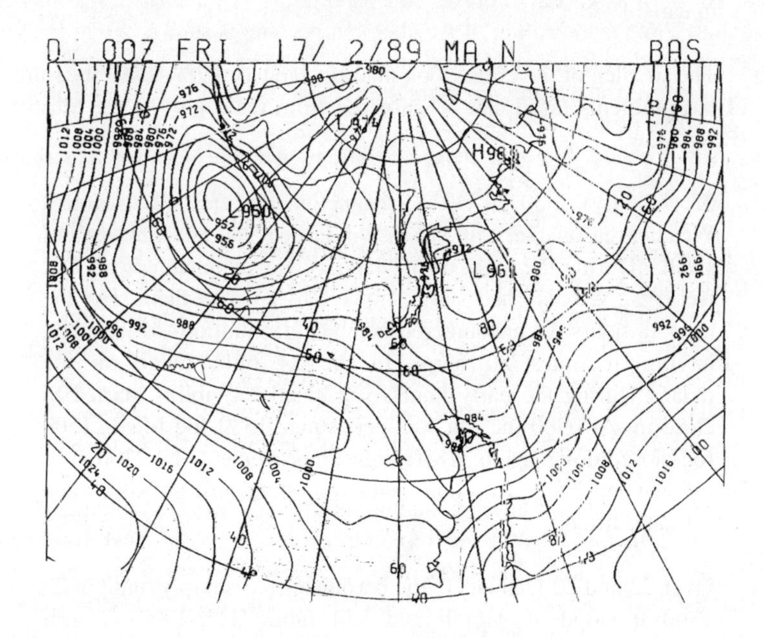

Plate 16: FAX of Antarctic weather.

was snowing, or if neither nearly always blowing. Gales frequently reached force 9, and one of them hurricane force for a period of about 8 hours. We also had to contend with large numbers of icebergs, growlers and bergy bits.

In this, RRS *Charles Darwin* appears to have followed its eponymous forebear, who noted in March 1834:

> The weather was very boisterous and cold, with heavy hailstorms; we got on, however, pretty well. Excepting some little geology, nothing could be less interesting.

In contrast, the second and third legs appear to have been relatively trouble free as regards weather, an interesting comment from Keith Avery's report being '....on Friday 10th (February) GLORIA, the hydrophones and the airgun were streamed and the ship rolled easily at 8 knots on an easterly course with wind rising to force 6-7'. Just how relative one's perception of weather can become is shown by the FAX reproduced as Black and White Plate 16, showing the ship on the rim of one area of intense low pressure surrounded by tightly packed isobars indicative of high wind speeds.

The contrast between these conditions and the ideal weather that every marine scientist seeks is shown by Colour Plates 5 and 7.

Well, for the scientists from BAS, 'some little geology' was why they came, and so they got on with it. However, this general pounding of the ship was always likely to cause some problems, more often than not irritating rather than major. One such problem showed that not everyone held scientific intelligence in high regard - the hot water urn in the duty mess developed a fault, and became unusable. Clearly feeling that he had to make this point abundantly clear to those on board who claimed a higher authority, the ship's electrical officer posted the notice shown on p127.

This message got through, and none suffered harm from the fault.

The principal result of the research for which the ship had to brave all this meteorological unpleasantness was four substantial surveys, the largest being 85,000 square kilometres of GLORIA sidescan sonar survey of the Antarctic Peninsula Pacific margin and adjacent ocean floor - the first time that such a survey had been made of any part of the

Antarctic continental margin (Tomlinson *et al* 1992). Charts of the area covered by the GLORIA surveys are shown on Colour plate 18. In its annual report for 1988-89 BAS (1989) notes:

> Over the northern part of the South Sandwich fore-arc GLORIA was used to look for evidence of serpentine diapirism similar to that recently reported from the Marianas fore-arc. The same survey extended across the trench to see if the tearing of oceanic lithosphere at depth, indicated by earthquake data in this region, has any expression at the ocean floor. Both searches proved negative, but demonstrating the absence of these features is in itself a significant contribution to studies of subduction-related processes and the mechanical properties of oceanic lithosphere.

thus demonstrating another aspect of scientific research - that data which do not support an hypothesis are as significant as data that do.

But the personal recollections of Carol Pudsey - watch leader on Leg 2, and joint Principal Scientist on Leg 3 of this cruise - are as interesting and evocative as the scientific results. Her record of a particularly poignant period in the cruise notes:

DO NOT SWITCH THIS APPLIANCE ON!
IT HAS BEEN SWITCHED OFF BECAUSE

It has a full earth!
No ohms between live and earth!
Not a sausage between the safe bit and the zig-zag bit!
Makes your hair stand on end when touched!
Ruins your whole day when touched!
Fifty very fast, mean, cyclists race through your body causing havoc!
Fills your underpants!
It doesn't like 'down-to-earth' guys!
It's not a pleasant experience!
You may damage the bulkhead when thrown across the duty-mess with the force of a Tyson slap-on-the-back!
The doctor's gone ashore for the day!

Saturday Feb 18th. South Sandwich survey area.

After sharing a bottle of wine with Peter (Dr Peter Barker, from BAS, the Principal Scientist for this leg) and Ric, I was sound asleep and dreaming when Peter woke me at 1930 and called me to a meeting with Phil and Rob. Captain Sam (Sam Mayl, who had been Master on a number of the ship's cruises, notably in Tonga) is no better and Dr Butler wants to get him ashore for a thorough examination without waiting 3 weeks till the end of the cruise. This will mean curtailing the South Sandwich survey, making a dash for Stanley, putting Sam ashore and dashing out again to lay the current meters. There won't be time for anything else.

Sam Mayl had flown from the UK with Keith Avery to become Master for the balance of the Antarctic cruise, but on the flight out had complained of a certain lethargy which he put down to recovering from flu. Once on board, his energy had seemed undiminished, and he had been an able Captain, and one of the Principal Scientists has commented that Sam's experience of Antarctic waters was beginning to result in a slightly more relaxed approach than that shown by the other two Masters on this cruise, who were noticeably more cautious about working in ice because it was new to them. But the evidence became more and more obvious that Sam was still suffering from the apparent after-effects of his 'flu'. He appeared less frequently on the bridge, and although his paperwork was, as always, impeccable, it was clear that something was wrong. By mid-February, it had finally become obvious that Sam Mayl was no longer in command of the ship, and the meeting referred to in the previous paragraph resulted, following a telephone exchange between Keith Avery and RVS Barry.

This concern is, perhaps, best expressed by Carol's entry in her diary a couple of days later, when she notes:

Monday February 20th South Sandwich Area

Went to visit Captain Sam about 9 o'clock. He was in bed though taking sustenance, and frankly I thought he looked terrible. The doctor says he has lost 1.5 stone, he's coughing and his voice is weak. He seems ten years older than the Sam I know. I'm not very good at making conversation and pretending that everything's fine when it obviously isn't. The sooner we get him ashore the better.

Keith Avery, who was second-in-command, noted a conflict between the requirements of ship-handling and science in these unfortunate circumstances. In his report to RVS, he wrote

Due to Captain Mayl's illness, the scientific gear was handed (brought inboard) at midnight 21st-22nd with the exception of the PES. The Senior Scientist requested that the magnetometer be streamed for the run to Port Stanley; I refused permission as I wanted the watchkeepers to have complete freedom of navigation in dangerous waters and as a watchkeeper myself I was rarely in a position to assist. When I went on the bridge at midnight ... I found a letter addressed to me in a sealed envelope from the Senior Scientist requesting that I reconsider my decision not to allow the towing of the magnetometer. I gave careful consideration to the written answer which I delivered by hand (I think) on Thursday 23rd.

This parenthetical comment is revealing in that in conversation with the writer, Keith said that he screwed up the letter from the PSO and threw it away in irritation, only later retrieving it for posterity.

Carol Pudsey continues

Monday Feb. 27th. Port Stanley - Falkland Plateau

Very impressed with Keith tonight. He took the ship in to Stanley Harbour and berthed at FIPASS[25] in the dark, and he's only been here once before. ...A memorable experience, coming in through the Narrows in the light of a bright half-moon, then seeing the lights of Stanley and FIPASS and smelling the peat smoke. Keith was on his own on the bridge for the approach, with Andy (Andy Louch, the Second Officer) on the bow and Jerry (Jeremy Clarke,

> the Third Officer) down aft. We tied up at 0215 so I finished my letter and took it and Sam's card up to the bridge to give to Keith and say some nice things. He said it felt a bit like driving an articulated lorry into a garage big enough for a Mini!

Keith himself has similar memories of this event, coloured of course by his concern for Sam as a colleague.

Sam Mayl was flown back to UK, and treated initially at the Princess Alexandra Hospital, at Wroughton near Swindon. A brain tumour was diagnosed, and he was transferred for treatment to the National Hospital for Nervous Diseases on 17 March. Peter Barker visited Sam there, and although weak, he continued to express interest in the progress of the cruise. Sam Mayl was later transferred back to Wroughton, where sadly he died on 4 April 1989.

However, Keith's memory also recalls the fulfilment of an early nautical ambition. When the time came to depart from Port Stanley, the wind was offshore. Thus Keith ordered the release of the lines aft then forward, and the wind gently eased the ship off the berth to the extent that he was able to ring for engine power to move the ship forward without need for manoeuvring.

Fortunately, an event such as the one above is unique (although illness or domestic problems can cause perturbation of a cruise), and Carol Pudsey continued to record the day-to-day minutiae of the remainder of the cruise. She notes in a letter that '...to convey what it's really like one should describe the monotony of being at sea for long periods!', but fortunately she does not dwell on that aspect. In mid-March, she notes:

> Saturday March 18th Drake Passage
>
> It blew an absolute hooligan (*sic*) from the west this morning, rumours of a 43° roll. All sorts of crashes from the galley at breakfast time .... Keith refused to stay on a SW'ly course and we have spent most of the day steaming slowly west at 4-5 knots in a truly enormous swell (estimated 40 ft). Every so often a wave broke over the bow and filled the foredeck with water and threw heavy spray over the wheelhouse windows (so

I got my heavy weather photos at last). ... All the gear (GLORIA, fish, magnetometer) was still out because it was far too dangerous to try to get it in. There wasn't much we could do except try to stay far enough off the wind to keep speed up (GLORIA is uncomfortable at less than 5 knots). I stood the last 1.5 hours of the afternoon watch then turned in for a while before supper. At 1730 I'd just had a plate of roast lamb set in front of me when Mac came along to say the GLORIA conducting cable had failed.

We do carry a spare but it cannot possibly be fixed under these conditions; calm water is needed, which means running for shelter in the South Shetland Islands. I gave the bridge a waypoint just NE of Smith Island and we can decide where to hide when we get there. Meanwhile the ship is behaving remarkably well in the circumstances. ... The galley have had a very hard time but have managed to produce nice food. ...

Monday March 20th Smith Island
(Smith Island bears the name of one William Smith, the Master of the brig *Williams* trading to South American ports. He discovered the South Shetland Islands whilst on passage between Montevideo and Valparaiso in 1819. He returned twice the following year to make good the discovery)

...We hove to about 4 miles SE of Smith Island; by evening it was almost cloud-free and looked stunning as always. I'm glad to have brought people here to show them a bit of the real Antarctic..

Tuesday March 21st Antarctic Peninsula Survey

Rob and Mac are both thoroughly twitched up about bad weather now and I keep having to pacify them.....This Principal Scientist's job is a strange mixture of prodding inert people into activity and calming hyperactive people down. So far, I'm rather taking to it.

This period covered the return from the Antarctic to Valparaiso, and once again one of the infrastructure items aboard showed signs of stress after the battering of the weather; this time it was the refrigerator in the Officers' and Scientists' Bar. It broke beyond repair, and had therefore to be removed. Andy Louch, the acting Chief Officer, records

> It was found that the fridge was too large to fit through the doorway. Appledore Shipbuilders must have installed the doorframe after installing the bar fittings. So a number of modifications were made with a hacksaw. Once the fridge would just fit through the doorway the Ship's Electrical Officer, Bosun and a sailor all supervised by the Second Engineer attempted to 'help' the unit in its passage to the outside deck. At a crucial stage, with the fridge just blocking the lobby, the ship took a wave down the port side. Fortunately all but the Second Engineer were inside and due to their portly physiques formed a perfect watertight seal in the doorway with the fridge. As the seawater was about 2 degrees Celsius the Engineer stranded outside leapt onto the spare lifejacket box and clung onto the lifeboat grab lines. His language at this sudden immersion was colourful and impressive once he had been able to draw breath following his icy shower. The human 'bung' prevented any ingress of water into the accommodation.

Such are the vicissitudes of life at sea.

The GLORIA survey ended at 62°S 64°W on 1 April, and RRS *Charles Darwin* headed north up the Chilean coast. All on board seized an opportunity to relax. Carol Pudsey continues

> Sunday April 9th Near Valparaiso
>
> Definitely a high-class punch at last night's party: I felt quite healthy this morning except for slight dehydration. Drinking my morning tea on the bridge with the third mate; he said it must have been a good party because he woke up still giggling ...

And so RRS *Charles Darwin* returned to Valparaiso, undamaged by her excursion into the unfamiliar and potentially hostile territory (if that is the appropriate word for a marine activity) of Antarctica, much to the relief of both her owners and crew.

[25] FIPASS - Falklands Interim Port and Stores System

12
Eastern Pacific Again

From Valparaiso, RRS *Charles Darwin* had sailed with Dr Brian Price as Principal Scientist to Balboa, in Panama. His discipline was palaeo-oceanography - the study of the geological history of the sediments and rocks that form the seabed. At the time of this cruise, the theories of Global Warming were being brought to public notice, and NERC publicity implied that Dr Price's cruise should provide evidence as to whether Global Warming was an anthropogenic phenomenon or a short period event in climatic cycles. The theory of Global Warming postulated a rise of some 10°C in average global temperature by the year 2030 - ie in about 40 years.

The formal remit of the cruise was:

> ..to undertake investigations in to the chemistry and sediments and water under areas of high coastal upwelling off the coast of Peru, and changing equatorial upwelling, and its history with time, on an equatorial section crossing the Equator west of the Galapagos Islands.

It was hoped that measurement of the biological material deposited on the seabed over time would provide some evidence of whether or not Global Warming was a modern problem. In fact, this was a publicist's wishful thinking. Dr Price pointed out that material from the ocean was deposited on the seabed at roughly one inch every thousand years, that this material was not only biological in origin, but also included substances introduced into the sea by, say, erosion by rivers or tectonic action. Thus the prospect of isolating evidence of cycle times as short as decades was so low as to be unrealistic.

In travelling from Valparaiso to Balboa, RRS *Charles Darwin* passed relatively close to the Galapagos Islands, on which Charles Darwin (the naturalist) carried out some of the studies on which his theory of evolution was based. This was of particular interest to the scientists on the cruise, as Charles Darwin had begun his academic career studying geology at the University of Edinburgh, which has a building named after him. Keynes notes 'Although Darwin had expected the main interest of the Islands to lie in their geology, it turned out otherwise...' One consequence of the attentions of the naturalist was the establishment by Ecuador of the Charles Darwin Research Station (CDRS) on the Islands.

As a result of these connections, RRS *Charles Darwin* visited the eponymous research station on Santa Cruz. The Director and staff welcomed this contact with their namesake, and described to the ship's scientists the conservation and biological research programmes being carried out by scientists from USA, South America, Japan and Northern Europe. This led to discussions about possible future co-operation on the assessment of El Niño effects in Galapagos waters.

Balboa sits at the southern (Pacific) end of the Panama Canal, and is one of the major cities in the State of Panama. It was where Cruise CD 38/89 ended and where CD39/89 was due to begin. Unfortunately, it was not only the weather which disrupted the careful logistical planning that had gone into the global circumnavigation of RRS *Charles Darwin* - politics also played their part. When RRS *Charles Darwin* berthed at Balboa, there was political unrest in Panama because General Noriega had cancelled the presidential elections due to be held at that time. The advice received from the UK Foreign and Commonwealth Office was that it was unwise to fly scientists into Balboa to join the ship; of course, an added incentive to accept this advice was that all flights into Panama had been stopped by the US carriers because of the problems! These problems led to two interesting anecdotes.

The first concerns the RVS staff who were due to meet the ship on its arrival in Balboa. Two of them, in the persons of Tony Cumming and

David Booth, flew out to Panama via Miami on the Sunday on which the Panamanian elections had taken place. Their flight from Miami was on a Peruvian Airlines plane, and at the time this airline was still flying to Balboa. They arrived at about 0200 on the Monday morning, and having cleared customs and immigration, looked for the taxi that they expected the agent to provide. No taxi had appeared by 0300, and a call to the agent produced the advice to wait a while. Still no taxi at 0400, and another call to the agent moved him to come to the airport in person. As they were about to leave the airport, an official asked to see their 'Canal Zone Clearance', which they did not have. This produced the comment that such clearance was necessary, and without it they were technically illegal immigrants in Panama. They were therefore hustled back to the airport side of the immigration barrier, and as no more planes were expected that night the immigration officials had gone off-duty, leaving only a caretaker and the prospect of an extended stay in official limbo. Tony asked the Agent to check on the 'cost' of obtaining the Canal Zone Clearances, and after some demurring a figure of US$20 was agreed!

On the journey from the airport into Panama City, Tony and David noticed groups of soldiery and civilians unhappy with the election results assembling in sidestreets. Closer to the centre, these groups had begun to build barricades, one of which impeded the final few yards to the hotel. Again, diplomacy was brought into play, and having noted that the travellers had UK passports, a further US$20 eased the opening of this barrier.

One member of the crew had flown in earlier, and had been booked into a different hotel. When he observed and reported that a Panamanian politician had been shot outside his bedroom window, it was decided that it would be preferable all round if all the RVS personnel stayed in the same hotel. Such are the joys of working for RVS!

Dr Martin Sinha and his team had flown to Miami to prepare for their cruise, because this was the most convenient airport from which to make the relatively short hop to Balboa. When they found that they would be unable to fly there, the telephone and FAX lines between their hotel, the University of Cambridge and RVS, Barry ran hot. It was eventually decided that RRS *Charles Darwin* would have to sail to San Diego and the cruise start from there.

No problem for those cosily ensconced in their offices back in UK to make this decision, but one major headache for the ship's staff. The logistics of the cruises required GLORIA to remain in Balboa during Sinha's cruise and to be loaded later for the following (Westbrook) cruise. Because of the political unrest it was deemed inadvisable to leave the system on the dockside. It was decided to load GLORIA prior to the ship's departure for San Diego. The dock staff were unable or unwilling to provide the manual and mechanical labour to effect this. The ship's Master at this time was one Captain 'Paddy' MacDermott, a pipe-smoking Irishman who, it was believed, had at one time kissed the Blarney Stone. Whether it was by his golden tongue or, as more widely accepted (but denied by Paddy), by the passing of folding greenbacks to a receptive crane-driver, he managed to overcome political problems, get GLORIA aboard and sail for San Diego, some 2,800 nm (or 11-12 days sailing) north-north-west.

Meanwhile, Martin Sinha discovered that DASI and its associated equipment, which was to have been transshipped from Cambridge to Balboa by container ship was, fortuitously, still in Miami because the vessel had suffered a collision in New York and was awaiting repair. By one of those curious co-incidences which make one believe that the gods occasionally smile on the righteous, Martin happened to have the Bill of Lading with him, and was thus able to recover his container from the dock on which it reposed with others from the ship. However, this was in Miami, and he needed the equipment in San Diego. Again, the gods smiled on Martin, and one of his fellow scientists from the Scripps Institute of Oceanography in La Jolla, just north of San Diego, just happened to know of a container lorry that was seeking a return load from Miami to San Diego. And so the equipment container travelled from Miami across Florida, Alabama, Mississipi, Louisiana, Texas, New Mexico, Arizona and California to San Diego at a lower cost than had been feared.

By this time, cost was a factor that was beginning to worry Dr Sinha, because his grant had been awarded on the basis of a relatively short stay in Miami for his team before they flew to Balboa to join the ship. Now they had to cross the continental United States to San Diego, which implied a worrying increase in both travel and accommodation costs. But young scientists, which the support team were, are notoriously flexible, and seven of them hired a 'Recreational Vehicle' and drove across from Miami, in one step minimising both travel and accommodation costs.

Meanwhile, RRS *Charles Darwin* arrived in the port of San Diego on a Saturday afternoon and US Customs and Immigration Services did not work over the weekend to clear the crew to come ashore. No problem, one would think, except that, because of the potential dangers implicit in the political uncertainty in Balboa, the crew had been kept aboard the ship at the completion of the previous cruise, and they had not had a run ashore for 43 days. Mutiny was in the air, but only because they were disgruntled at being cooped up for 6 weeks without the opportunity to stretch their legs (or whatever else it is that sailors do ashore). Again, Martin Sinha's contacts at Scripps came to the rescue, and a senior officer in the Customs and Immigration Service was persuaded to make a special trip to clear the ship, so that the crew could get their run ashore.

And so, on 28 May, some two weeks later than planned, Dr Martin Sinha's team sailed on RRS *Charles Darwin* Cruise CD 39/89 to begin their study of the East Pacific Rise. Sadly for science, the planned working area, between 9°N and 11°N was too far south of San Diego to be reached in the time available - a cardinal rule enforced by RVS was that cruises had to finish on their planned date: no time was allowed for unforeseen problems, in order not to disrupt the timetable for subsequent cruises[26] - and it was decided to work at 13° 15°N, another area which previously-obtained data indicated would be of scientific relevance.

The science required the deployment of acoustic navigation transponders and LEMURS (**L**ow-frequency **E**lectro-**M**agnetic **U**nderwater **R**ecorders, which are free-fall instruments deployed on the seabed at an appropriate range - typically a few kilometres - from the transmitter and used to record the resultant oscillating electric fields (Sinha *et al*, 1990)) on the sea bottom, the first to provide a basis for extremely accurate navigation, the latter as receptors and recorders of the electro-magnetic signals generated by DASI. In the event, both deployments were effected satisfactorily, DASI worked without a hitch, and Martin Sinha was able to report (Sinha & Constable 1989):

> With the aid of the acoustic navigation system it proved possible to keep the deep tow (DASI) within 200 m of planned tracks all of the time and within 50 m much of the time. Flying height was maintained at 30 ± 10m.

The outstanding piece of equipment on this cruise was undoubtedly the Oceano acoustic navigation system, without which we could not have performed the experiment nearly so well. All the electromagnetic sounding equipment worked well, and clear active source signals have been observed on all four of the Scripps receivers and one Cambridge instrument from which data have so far been replayed. The source exceeded expectations, and demonstrated clearly the viability of flying a 180 m long neutrally buoyant transmission antenna very close to a rough seafloor along precisely controlled tracks. ...

Despite an unpromising start, this cruise has been a tremendous success. Recording of 32 hours of transmitted data on and off the ridge axis by an array of seven seabottom instruments has produced a superb data set. Although in the reduced time we were not able to extend our work as far off-axis as we had planned (phase 2), we have fully achieved the objectives of the original phase 1.

Both Britain and the US are planning major initiatives to study mid-ocean ridge processes. In this experiment we have demonstrated that despite the novelty and technical difficulty of the seafloor active source EM method, extensive and detailed surveys of ridge-axis areas are possible. Given the special sensitivity of the EM method to melt content, porosity and temperature, the availability of this new technique will greatly enhance the results that could be obtained using conventional geophysical approaches alone.

A subsequent Press Release by NERC[27] put the scientific findings in a much more flowery way. Headed '*Probing the planet's 'crust creators*'', it said:

Scientists have used a new electromagnetic technique to probe deep into the chambers of molten rock (magma) below the seabed. These magma chambers play a vital role in the creation of the Earth's crust, provide the source for volcanic eruptions, and are linked with the existence of unique mineral-rich 'hot springs' on the seabed, where life flourishes without sunlight.

'Crustal magma chambers are the key to the geology of two-thirds of the Earth's surface' said Dr Martin Sinha, who led the two research cruises. 'We are learning more about the size of the chambers, how much of the material in them is liquid and how much crystallised, and how long it takes for the magma to cool and form new crust.' The research from RRS *Charles Darwin* suggests that magma chambers are replenished every few thousand years by pulses of melt from the mantle of rocks below the Earth's crust.

In contrast to all the excitement with Dr Sinha's cruise arrangements, Graham Westbrook knew about one month in advance that his cruise would begin in San Diego, rather than Balboa. Nevertheless, he was also upset by the fact that the change of port involved himself and the ship in 11 days of transit time that would otherwise have been spent on science. He faced the added problems that he had to pick up a Colombian Navy observer from Tumaco (in the south of Colombia) before research could begin in Colombian waters, and then collect from Balboa the multi-channel seismic equipment which would not be in San Diego before the ship sailed. But even the picking-up of the Colombian observer was not without its drama. The location of the observer was eventually determined by telephone conversations between the British Consulate in Cartagena, Chris Adams (RVS' Operations Officer) at home and satellite communication with the ship. It transpired that the observer was in a boat in a bay on the west coast of Colombia and Keith Avery had to find the precise spot on his charts to pick him up. That he did is a tribute to the power of modern communications as well as to Keith's navigational skills.

The formal remit of Professor Westbrook's cruise was:

> ..to undertake a study of the continental margin of SE Panama, Colombia and NW Ecuador, to understand how the subduction of ridges, rifts and other features of the ocean floor, including variation in sediment thickness, control whether sediment is

Plate 17: GLORIA montage from Westbrook's cruise CD40.

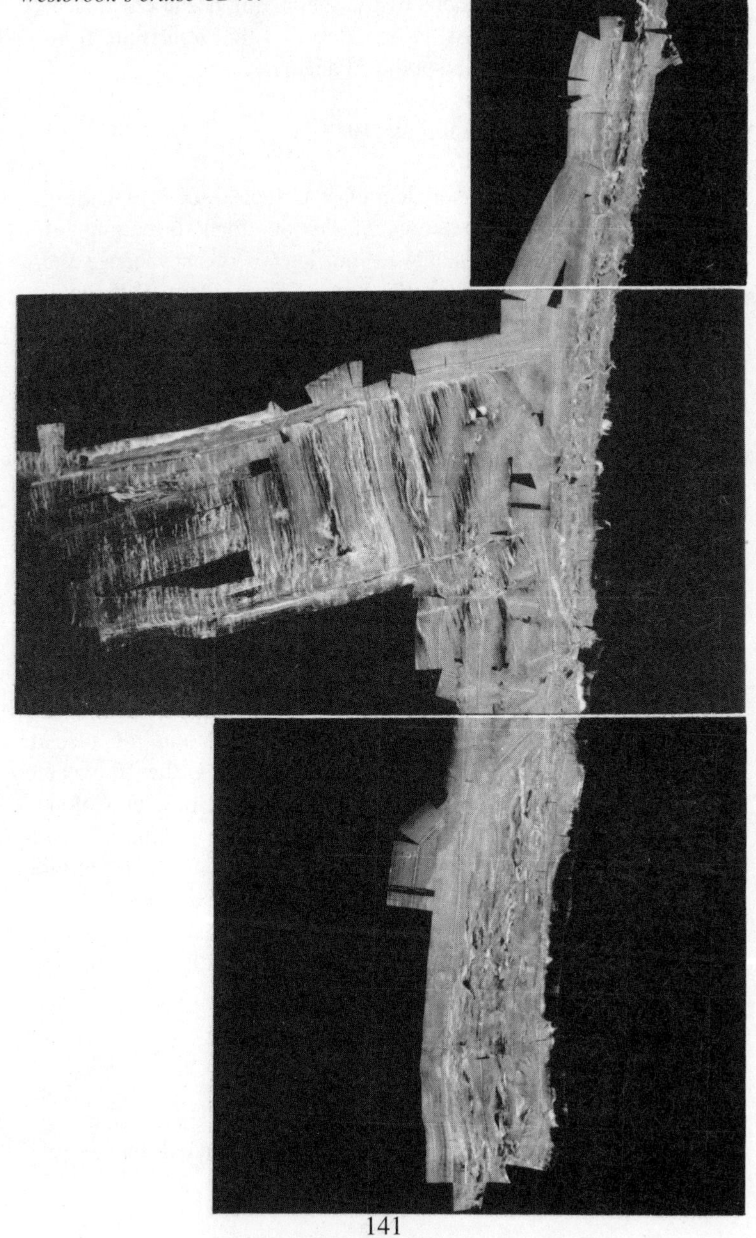

accreted to the margin, or whether it is subducted and, if the margin is tectonically eroded by the subducting ocean crust. Also important is the change in structure in the transition from convergence to strike-slip south of Panama.

In his Cruise Report Westbrook (1989) writes:

> The cruise, despite its short length, was successful in obtaining data that reveal new features of the continental margins of Colombia and Panama that were not known of and change the currently accepted ideas on the type of convergent continental margin that is present.
>
> The seismic data...together with the gravity data...indicate that the whole of the margin is composed of deformed sediments, both accreted sediments and sediments in fore-arc basins...The top of the Nazca Plate can be traced landward a few tens of km beneath the accreted sediments...
>
> The edge of the continental margin (accretionary complex) was clearly shown in the GLORIA sonographs, and it varies greatly in style. Some portions have broad gentle folds forming, others are more variable with intenser deformation...

In his final (confidential) report on the outcome of research supported by his NERC grant Westbrook notes the unexpected revelation of a 40 km shelf off the northernmost part of Colombia 'where data available prior to the cruise had led one to expect a margin dominated by tectonic erosion, and with a trench that had no turbidite fill where a substantial thickness had been expected'. These conclusions, coupled with the more quantitative analyses of data that provided information on what had been expected led Prof Westbrook to suggest a number of new research activities, viz:

> The fluid-flow regime of the accretionary complex requires further investigation...New data specifically orientated toward the detection and measurement of fluid flow are required...deep-tow sidescan sonar should be used to provide detailed images of

> potential vents, which can be followed up with measurements of fluid flow made with pore pressure instruments, and with flow meters and water samplers deployed from a submersible or ROV...How the transform fault south of the Gulf of Panama translates its motion into the zone of oblique convergence off south west Panama is an interesting problem that can be investigated with multibeam bathymetry, sidescan sonar and seismic reflection...

Thus does the interest and intellect of science develop, when a project produces unexpected results and stimulates questions about new opportunities revealed by the opening of a new door. But these conclusions were reached, in part, because the data gathered on the cruise stimulated the development of new quantitative analytical techniques. As an example of the support functions that complement the (relative) glory of an actual cruise, Westbrook's own words tell the story.

> A major development has been the production of software for processing GLORIA data on UNIX-based computer workstations... An important innovation was to use a digital terrain model of the seabed, constructed from bathymetric data from the cruise and all other available bathymetric data in the area, to derive the variation in response of the sonar to range, depth, and angle of slope of the seabed. This response function, obtained with interpolation and smoothing from the response of every pixel of the sidescan data set, was inverted to derive a correction function that will leave the sidescan sonar image as close as possible to a map of the reflectivity of the seabed. The efficiency of this approach was demonstrated strikingly on survey lines running parallel to the continental slope, where regions imaged from both downslope and upslope have different overall brightness when corrected using the approach adopted in the USGS[28] MIPS software, producing a sharp boundary where the upslope and downslope sonographs are joined together. This problem is alleviated using the Birmingham software so that the region has the same brightness irrespective of whether it is insonified from upslope or downslope. The interpretation of the sonographs has been aided by the development of a facility for on-screen annotation and marking of features, so that the boundaries can be drawn to produce maps that are displayed in overlay form on the sonographs or separately. In the interpretation of the 3.5 kHz

records a method has been developed in which the data are coded according to the characteristics of the records such as whether the seabed is locally rough or smooth, whether the subseabed is acoustically opaque or transparent, layered etc. These characteristics are displayed in colour-codes along the tracklines on the sonograph to aid the correlation between seabed reflectivity and the nature of the seabed and its immediate subseabed character.

This extended quotation is included to highlight for interest some of the major changes that science has undergone in the 150 years since Mr Charles Darwin sailed on HMS *Beagle* in this part of the world. His geological research equipment comprised essentially a geological hammer, a notebook, and a keen sense of observation. Whilst all three remain essential requisites for the geologist - both amateur and professional - Graham Westbrook shows that the modern marine geophysicist needs not only sophisticated (and costly) data-systems on board ship, but also analytical and computational support quite undreamed of by the naturalist. Indeed, the quotation itself shows that analytical techniques developed over only the past 10-15 years have now been superseded by methods that recognise improved data-gathering and computational power. By such synergy of scientific push and technological pull does science advance.

RRS *Charles Darwin* completed her activities in the Pacific Ocean and returned to Balboa, and the Panama Canal. In marked contrast to the problems that had caused the major logistics changes to Martin Sinha's cruise, the passage was uneventful, and the Master's report notes 'Including the time at anchor, the transit took 15.4 hours'.

---

[26] Although this may seem somewhat arbitrary and harsh, the diplomatic clearance requirements for cruises specify precise dates of operation. If those dates were changed at too short a notice, then the nation or state in whose waters the cruise was to take place would be within its rights to refuse clearance for that cruise, and later cruises would also be affected.

[27] *NERC News Release* 19/90. 19 June 1990. *'New Horizons in the Oceans - The RRS Charles Darwin Global Voyage of Discovery'*

[28] United States Geological Survey

13
Back into the Atlantic

Interest in the continental margin off Colombia continued on the eastern side of Panama, and the ship put in to Cartagena to disembark the current scientists and to pick up a team of American scientists, led by Dr Nancy Breen, from the Lamont Doherty Geological Observatory (LDGO), New York. Her particular interest was the effect the weight of sediment in the rapidly-growing Magdalena submarine fan would have on the (reputedly) active accretionary wedge in the margin. Or put more formally, she wished to:

> ..undertake a mapping study of sedimentary structures on the northern Colombian convergent margin and the Magdalena Fan, in order to study the effect of deposition of a large sediment load on the convergent margin structural geometry.

The origins of this research can be traced back to personal contact between Graham Westbrook and Nancy Breen, who had participated in a joint cruise on the American research ship R/V *Conrad* off Barbados in 1988. Nancy was a very accomplished women's rugby player, who had had her photograph on the cover of an American rugby magazine. At the time of the cruise, the British Lions tour of Australia was underway, and Nancy would hold long and heated conversations with Huw Evans (one of the NERC technicians, and a keen fan of Welsh Rugby Union) about whether the tactics employed by the Lions were right or not. Huw and his colleagues had to overcome their inbuilt prejudices to accept that a *woman*, and an *American woman* at that, could know *anything* about rugby, let alone sufficient to argue with them. Eventually they were convinced!

Dr Breen has moved at least twice within the USA between the time of her cruise and the writing of this book, and it proved difficult to track her down. The reference list includes papers indexed in the LDGO library that she has produced using data gathered on this cruise. However, in her cruise report, she notes:

> One of the main questions we had originally posed is: Where is the deformation front? Locating the position and geometry of the active deformation is essential in understanding the mechanical effect of the load of the Magdalena Fan on the convergent margin deformation. The absence of convergent deformation on the middle Magdalena Fan suggests that the fan is at a sub-critical taper, and has suppressed sliding on the decollement. Further support for this idea is given by folding deep within the sedimentary section on the eastern edge of the fan. This folding dies out upsection and is covered by flat-lying sediments. The implication is that deformation was formerly active but was suppressed by the sedimentation, and the position of the active deformation has moved towards the continental margin. Where we could map a change in the position of the thrust front, it appeared to step, rather than curve, out into the basic sediments on either side of the fan.

In other words, this survey appeared to indicate that the sheer weight and disposition of sediment can affect the physical structure of new material formed at the boundary between two tectonic plates that are moving together. And with this conclusion we reach the end of the geophysical research activities supported by RRS *Charles Darwin* over a three-year period. We have seen that, although the topic embraces a timescale measured in millions of years, much of the development of the science has occurred within the lifetime of the scientists carrying out the research, and those whose research was carried out from the ship have added to, and stimulated, the further development of current theories.

At the conclusion of the geophysical survey of the Magdalena Fan, RRS *Charles Darwin* headed for the most convenient port, in this case Kingston in Jamaica, where the scientific team were disembarked, the Americans to fly back to New York, the British to England. And then the ship herself steamed up the eastern seaboard of the United States, bound for New England. This passage was uneventful, although overshadowed by the threat of Hurricane '*Dean*'. The Master's report notes:

> Shortly after leaving Kingston, the forecast was that Hurricane '*Dean*' would be passing our track. To avoid this, direct drive was engaged which gave us an extra knot on our speed, just enough to keep us out of the danger area. Having got ahead, '*Dean*' swung to the north and ran on a more-or-less parallel track, being a threat to us all of the time. Local radio stations in the US described the course as unpredictable, so my reasoning was that the further we were away, the better. The direct drive was kept engaged until we were safe from '*Dean*' then we reverted to normal electric drive for the rest of the run. For the last couple of days the visibility slowly deteriorated until the final night we ran into two banks of dense fog. We cleared the second as we approached Woods Hole, and shortly after we docked at 0808 the fog came down very thick for the rest of the morning.

Woods Hole is a relatively small, but delightful, New England harbour with a regular ferry service to Martha's Vineyard. Once a major fishing port for the Grand Banks, it is today dominated by the Woods Hole Oceanographic Institution (WHOI), whose laboratories, originally based around the port area, now spread throughout the town. It was to be the last port of call by RRS *Charles Darwin* prior to her return to her home port. Because of this the crew had made special efforts on the voyage north to ensure that the ship's paintwork was in as good a condition as possible, but just to be sure that all would be well on her return, Captain Mike Perry, the Marine Superintendent from RVS flew over to take a personal look.

After two days in port, restocking logistically and preparing to sail with a team of scientists from WHOI, the ship left Woods Hole on 10 August for the north west Atlantic for the recovery of 18 moorings and the redeployment of six of them as part of the SYNOP (**SYN**optic **O**cean **P**rediction) experiment (Tarbell *et al* 1992). The success of this cruise is recorded in the report by one of the Principal Scientists (Worrilow 1989):

> Between the period of 13 and 25 August a total of twenty-four mooring operations were completed and seventeen CTD stations

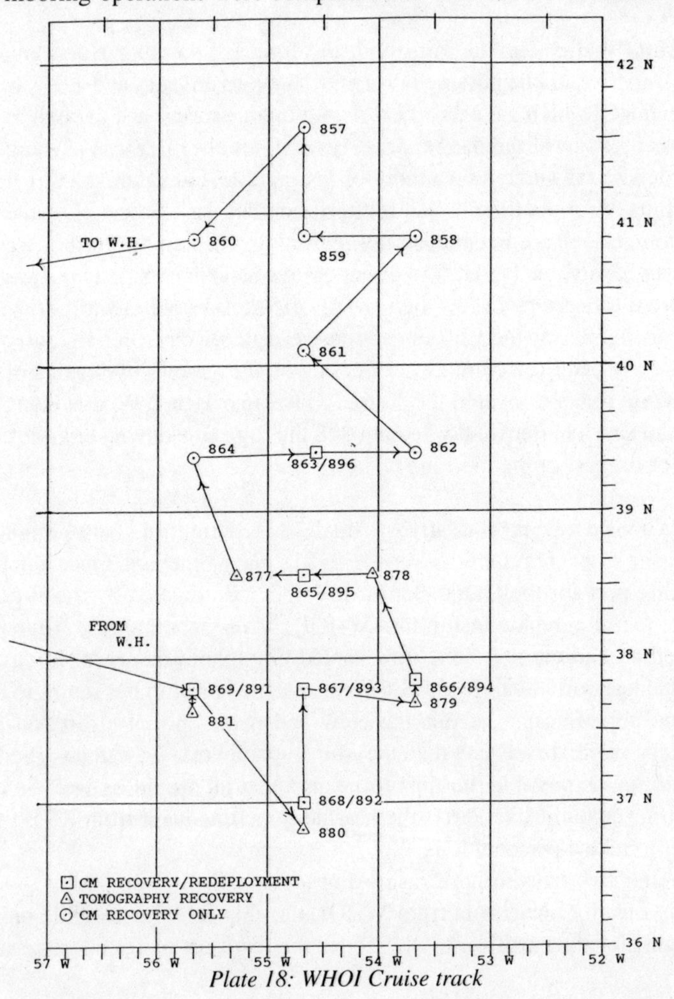

Plate 18: WHOI Cruise track

were taken. Because the weather, for the most part, was very good, the mooring operations went very smoothly, allowing us to finish the mooring work and return to Woods Hole on 28 August at 1500 hrs.

This quiet reporting conceals a deep sense of achievement by both the scientific team and the ship - the clean sweep of a completely successful recovery and deployment of the suite of moorings had only been achieved by WHOI twice in the previous decade. As a consequence, RRS *Charles Darwin* re-entered Woods Hole carrying a broom at her masthead.

Hogg (1991) used the data from these moorings to develop improved techniques for correcting data for the effect of flow on the depth of the recording instrument - ie a mooring string will move in the current it is designed to measure, and instruments will not necessarily be measuring parameters at their intended depths. Hogg further (1992, 1994 in press) used the data to model the eddies and meanders of the Gulf Stream in this area, whilst the data themselves were made widely available by Tarbell *et al* (1992).

Susan Tarbell was one of the WHOI team for this cruise. Her speciality is data processing, hence the citation immediately above. Her hobbies are photography and needle-point, the former providing some of the pictures reproduced in the Colour Plates, and the latter leading to an interesting story. She designs her own pictures, and on this cruise decided to represent the ship as a sea-dragon. Dragons have a certain reputation for fire-breathing, and the good ship RRS *Charles Darwin* was to be no exception. The only problem was that Susan's representational flames were keenly examined by the professional eyes of the engineers on board, who proclaimed themselves dissatisfied with the portrayal, and required Susan to unpick a significant part of her picture and insert more appropriately designed fires! Of course, such attention to pictorial correctness on the part of the engineers had no connection with the fact that the ship's home base was in Wales, whose national flag consists of a red dragon on a field of green and white!

But this cruise also had a unique poignancy. In Chapter 11, the illness and death of Captain Sam Mayl was recorded. His last wish had been to have his ashes committed to the seas which were such an important part of his life, and so Mike Perry had brought Sam's ashes with him to Boston. These were taken to sea on this cruise, and the ship's Master (Paddy MacDermott) performed a simple but moving service of committal, and Sam Mayl became one with the Atlantic Ocean. Whilst

The Master, Officers and Crew of the R.R.S. Charles Darwin would like to convey their heartfelt thanks to the Woods Hole Oceanographic Institution and in particular the Buoy Group, Public Relations Office and Dan and Mavis Clarke for their unstinting and gracious hospitality shown during her recent visit, it was unparalleled.

P.J. MACDERMOTT MASTER.

Plate 19: Copy of Paddy's certificate

burial of the dead at sea is catered for in the appropriate manual of seamanship, the committal of ashes is not, and it was a curious coincidence that Paddy had already conducted such a ceremony when Master of RRS *Challenger*.

When RRS *Charles Darwin* returned to Woods Hole two days earlier than projected, two things became obvious - firstly with no more science to carry out before her return to UK, an early departure would result in an embarrassingly early arrival, and disrupt the timing of the events being planned there to mark the completion of the voyage. The second was the innate concern of the Master and crew that the ship should look her best on return, showing as few scars as possible after 3.5 years travel. These two concerns worked together to encourage all concerned to make full use of the 2-3 days in Woods Hole to get RRS *Charles Darwin* all 'ship-shape and Bristol fashion'.

But of course, nowhere is more conducive to fulfilment of the aphorism 'all work and no play makes Jack a dull boy' than being among like-minded marine and scientific colleagues in the USA. The ship's agents in Woods Hole were Dan and Mavis Clark, an erstwhile English couple but longtime Woods Hole residents and friends of WHOI (Dan, a Naval Architect, developed the original support vessel (*Lulu*) for the deepsea submersible *(Alvin)*), and so given the presence of a UK ship in Woods Hole with time to spare they were naturally included when the WHOI Physical Oceanography and Buoy Groups organised a barbecue in the grounds of WHOI. All reports suggest that it was quite a party but, apart from those who forged new and subsequently long-lasting friendships, memories appear somewhat hazy on detail! That there was something different about this visit and its associated hospitality is shown by the fact that it is only for this cruise that there remains a formal scroll from the ship, signed by the Master, shown opposite.

Some 5 years after the event, Mavis Clark recalls the visit by RRS *Charles Darwin* with fondness and enthusiasm, and one or two happenings that were a mite unusual. The first concerns the ship's laundry: after an extended period of scientific cruises with little opportunity to use effective port agencies to bring the laundry up-to-date, the ship was in especial need of clean sheets and other linen. As noted above, Woods Hole is essentially a village, and as such does not possess an industrial or commercial laundry service, so the Clarks organised teams of local ladies to use launderettes and their own machines to wash and iron the ship's linen. Mavis also remembers Mike Perry's predilection for the large American automobiles - the 'stretched limos' - and tells with delight of Mike's face when, for his return to Boston they arranged for him to travel the 90-odd miles in one such limo, complete with TV and cocktail cabinet in the back. (And before the bean-counters seek to recover the cost of this apparent extravagance from someone, Mavis and Dan Clark paid for it themselves as a thank you for news and company of those from the land of their birth).

But let us close this chapter with a recollection from Paddy MacDermott. He, too, recalls with fondness the Woods Hole portcall, and speaks of Dan Clark as a big man, not only because of his six feet plus height, but because of the genuine warmth of his friendship. It apparently became something of a tradition (if such can be built up in a few days) for Paddy to visit Dan's house each evening, when the whiskey cupboard would be opened, and unusual and rare brands sampled with accompanying puffing on pipes and wide-ranging conversation. And then, when Paddy was up to his ears on the morning of departure and almost ready to utter those famous words 'Let go aft', Dan appeared on the dockside with a *'bon voyage'* gift which was savoured and enjoyed on many a palate long after the cruise was over. Such was the rapport that Paddy maintains that if it had been any caller other than Dan at that particular juncture in the departure process, he would have denied him access.

14
Homecoming

At midday on 2 September 1989, RRS *Charles Darwin* left Woods Hole on the final voyage of her global circumnavigation, to RVS Barry and home (see Colour Plate 13). Even this trip was not to be without incident - another hurricane ('*Felix*') was disturbing the relative calm of the north Atlantic, and the ship was forced to divert from the most direct course to minimise potential problems. However, early in the evening of Friday 15 September, RRS *Charles Darwin* came through the lock at the entrance to Barry Docks, and the harbour pilot manoeuvred her alongside to the cheers of the waiting Base staff and the (muted) glare of publicity from press and broadcasting media.

'World trip scientists head for home' was the headline in the *Western Mail*; '*Study on mystery'* claimed the *Glasgow Evening Times*; '*Weather riddle in focus'* declared the *Liverpool Echo*; '*Scientists uncover natural source of greenhouse gas'* stated *The Scotsman*; whilst *New Scientist* prosaically summed up the whole event with '*Charles Darwin comes back home with a cargo of data'*. Given the contrast implicit in a (probably) once-in-a-lifetime event in one branch of the environmental sciences, the coverage of the return was gratifying.

Of course, the press articles were themselves based upon a NERC Press Release, from which the various headlines were developed, but it was left to schoolchildren to ask the really important questions. Following a relatively quiet weekend after the ship's return, parties of schoolchildren were invited to visit the ship. They were given fact sheets about the ship herself and about the environmental sciences that she had supported for the past three years, and the children would be asked questions at the conclusion of their visit. They were also encouraged to ask questions during the tours, and the very first one floored all the support staff conducting the tours - 'Please sir, how many miles per gallon did the ship do?!' (9 gallons per nautical mile.) This caused the rapid reproduction of a revised facts sheet (shown at Appendix 2), with honour satisfied all round.

To mark the return more formally, NERC's Chairman, Professor John Knill, hosted a reception on board the ship for senior representatives of many of the countries and scientific organisations which had been involved in the ship's cruises. This was followed by a seminar, held at the University of Wales College of Cardiff, at which speakers outlined the political, logistical and scientific aspects of the circumnavigation[29]. The programme for this seminar is shown at Appendix 3.

And so as RRS *Charles Darwin* ends its voyage, how does this compare with the ending of the voyage by the man himself? In his diary on his return, Charles Darwin wrote the following cautionary, not to say prophetic, words:

> Our voyage having come to an end, I will take a short retrospect of the advantages & disadvantages, the pains and the pleasures, of our five years' wandering. If a person should ask my advice before undertaking a long voyage, my answer would depend upon his possessing a decided taste for some branch of knowledge which could by such means be acquired. No doubt it is a high satisfaction to behold various countries, & the many races of Mankind, but the pleasures gained at the time do not counterbalance the evils. It is necessary to look forward to a harvest however distant it may be, when some fruit will be reaped, some good effected.

Whether, at this time (1836) Charles Darwin foresaw his *magnum opus* as the 'fruit' of this voyage can only be guessed at, whereas it is certain that every one of the scientific teams that sailed in the eponymous modern vessel had clearly in mind that the end-result had to be scientific papers aimed at enriching their particular disciplines - undoubtedly possessing a decided taste for some branch of knowledge. As the geophysicists, the marine physicists and chemists, the palaeogeologists and biogeochemists of today will affirm, it was that thirst for knowledge, an enrichment of their topic, that took them to the far corners of the world's oceans. It was that anticipated thrill, mentioned in connection with Rana Fine's discovery of a hitherto unknown current in the Indian Ocean, that fired each and every one to overcome whatever problems there were - political, financial, technical - to produce results.

This record has endeavoured to set down more than simply the purely scientific aspects of the global circumnavigation: let it end with a further note from the great man himself, a comment which may, perhaps, commend itself to his successors in science.

> But I too have too deeply enjoyed the voyage not to recommend to any naturalist to take all chances, & to start on travels by land if possible, if otherwise, on a long voyage. He may feel assured he will meet with no difficulties or dangers (excepting in rare cases) nearly so bad, as before hand imagined. In a moral point of view, the effect ought to be to teach him good humoured patience, unselfishness, the habit of acting for himself, & of making the best of everything, or contentment: in short, he should partake of the characteristic qualities of the greater number of sailors. Travelling ought also to teach him to distrust others; but at the same time he will discover how many truly good-natured people there are, with whom he never before had, nor ever again will have any further communication, yet who are ready to offer him the most disinterested assistance.

---

[29] Frank Verdon, 'New Horizons in the Ocean', *NERC News* Vol./Issue (October 1989): p12

Postscript

In 1990, NERC celebrated the 25th Anniversary of the granting of its Royal Charter. To mark the event, it staged a symposium jointly with the Royal Geographic Society, entitled 'New Horizons in the Oceans', which was devoted to science supported during the global voyage of RRS Charles Darwin, and many of the presentations were subsequently reported in Marine Geology 104. A more populist article was printed in the journal of the RGS[30]*. The ship herself was brought into the Pool of London for the two days of the symposium, and again played host to a number of receptions for political and scientific representatives.*

---

[30] *Frank Verdon, 'Fathoming the Seas', Geographical Magazine Vol/Issue (January 1990): 32-35*

Glossary

ADCP -	Acoustic Doppler Current Profiler
BAS -	British Antarctic Survey
BOFS -	Biogeochemical Ocean Flux Study
CFC -	Chloro-Fluoro-Carbon
CTD -	Conductivity, Temperature and Depth (sensor)
CZCS -	Coastal Zone Colour Scanner
DASI -	Deep-towed Active Source Instrument
FCO -	Foreign and Commonwealth Office
GLORIA -	Geological Long Range Inclined Asdic
IMER -	Institute for Marine Environmental Research
IOSDL -	Institute of Oceanographic Sciences - Deacon Laboratory
JGOFS -	Joint Geochemical Ocean Flux Study
LTHE -	Low Temperature Heat Exchanger
MCDSAS -	Multi-Channel Digital Seismic Acquisition System
NERC -	Natural Environment Research Council
ODP -	Ocean Drilling Program
PES -	Precision Echo Sounder
PML -	Plymouth Marine Laboratory
RRS -	Royal Research Ship
RSMAS -	Rosenthiel School of Marine and Atmospheric Sciences
RVS -	Research Vessel Services
RVSC -	Research Vessels Strategy Committee
UCNW -	University College, North Wales
UOR -	Undulating Oceanographic Recorder
WHOI -	Woods Hole Oceanographic Institution
WOCE -	World Ocean Circulation Experiment
XBT -	eXpendable Bathy-Thermosalinograph

References

(**Note** - not all the references given below are quoted in the text. However, all the scientific papers listed resulted from cruises undertaken on RRS *Charles Darwin*)

Ayub A (1992) Channel development and morphology on the Indus submarine fan. PhD Thesis. University of Wales

Barker PF & **Pudsey CJ** (1991) Cruise Report RRS *Charles Darwin* Cruise CD 37. January to April 1989. *British Antarctic Survey Report* 42pp

Barton ED & **Hill AE** (1990) Abyssal flow through the Amirante Trench (western Indian Ocean). *Deep Sea Research* Vol **36** No 7, pp 1121-1126

Bauer S (1990) Influence of monsoonally-forced Ekman dynamics upon surface layer depth and plankton biomass distribution in the Arabian Sea. MS Thesis. University of Miami.

Bauer S, Hitchcock GL, & **Olson DB** (1991) Influence of monsoonally-forced Ekman dynamics upon surface layer depth and plankton biomass distribution in the Arabian Sea. *Deep Sea Research* Vol **38**, No 5 pp 531-553

Bauer S, Hitchcock GL, & Olson DB (1992) Response of the Arabian Sea surface layer to monsoon forcing. *Oceanography of the Indian Ocean.* pp 659-672

Baxter A (1987) Cruise Report RRS *Charles Darwin* 21/87 11 March - 7 April 1987. City of London Polytechnic, 23pp

Beekman F, Bull JM, Cloetingh S, & Scrutton RA (1994 in press) Crustal fault re-activation as initiator of lithospheric folding in the Central Indian Ocean. *Geology.*

Bertram CJ (1989) Rare earth elements and neodymium isotopes in the Indian Ocean. PhD Thesis. University of Cambridge

Binard N, Hekinian R, Cheminee JL, Searle RC, & Stoffers P (1991) Morphological and structural studies of the Society and Austral hotspot regions in the South Pacific. *Tectonophysics.* **186** pp 293-312

British Antarctic Survey (1989) *Annual Report of the British Antarctic Survey for 1988-1989* Natural Environment Research Council pp 48-49

Bull JM (1990) Structural style of intra-plate deformation, Central Indian Ocean Basin: evidence for the role of fracture zones. *Tectonophysics* **184**.pp 213-228

Bull JM (1990) The structural style of intraplate deformation, Central Indian Ocean Basin. PhD Thesis. University of Edinburgh. 218pp

Bull JM & Scrutton RA (1990a) Fault reactivation in the central Indian Ocean and the rheology of the oceanic lithosphere. *Nature* Vol **344**, Apr 1990. pp 855-858

Bull JM & Scrutton RA (1990b) Sediment velocities and deep structure from wide-angle reflection data around Leg 116 sites. *Proceedings of the Ocean Drilling Program, Scientific Results.* Vol **116**. pp 311-316

Bull JM & **Scrutton RA** (1992) Seismic reflection images of intraplate deformation, central Indian Ocean, and their tectonic significance. *J. Geol. Soc*, London. Vol **149**. pp 955-966

Bull JM, Martinod J & **Davy P** (1992) Buckling of the oceanic lithosphere from geophysical data and experiments. *Tectonics*. Vol **11**, No 3. pp 537-548

Burkill PH, Mantoura RFC & **Owens NJP** (Guest Editors) (1993) Biogeochemical Cycling in the north-western Indian Ocean. *Deep Sea Research II* Volume **40** No 3, pp 643-849

Burrows M (1994) Water and sediment movement in the east Ionian Sea. MSc Thesis. University of Southampton.

Clark JD, Kenyon NH & **Pickering KT** (1992) Quantitative analysis of the geometry of submarine channels: Implications for classification of submarine fans. *Geology* **20**, pp 633-636

Collier JS (1990) Seismic imaging of magma chambers and mud diapirs. PhD Thesis, University of Cambridge

Collier JS & **Sinha MC** (1990) Seismic images of a magma chamber beneath the Lau Basin back-arc spreading centre. *Nature*. **346**. pp 646-648

Collier JS & **Sinha MC** (1992) The Valu Fa Ridge: the pattern of volcanic activity at a back-arc spreading centre. *Marine Geology*. **104** pp 243-263

Collier JS & **Sinha MC** (1992) Seismic mapping of a magma chamber beneath the Valu Fa Ridge, Lau Basin. *J. Geophys. Res*, **97**. pp 14,031-14,053

Collier JS & **White RS** (1990) Mud diapirism within Indus fan sediments: Murray Ridge, Gulf of Oman. *Geophys. J. Int*. **101**, pp 345-353

Cook MF, *et al* (1992) A Trans-Indian Ocean hydrographic hection at Latitude 32°S: Data Report of RRS *Charles Darwin* Cruise #29. Woods Hole Oceanographic Institution *Technical Report WHOI* -92-07. 190pp

Cramp A (1987) RRS *Charles Darwin* Cruise 27/87 Cruise Report. Department of Earth Sciences, Swansea and Institute of Oceanographic Sciences, Wormley. 17pp

Cramp A & **Postma G** (1994 in press) Turbidite and debris flow deposits within ponded basins: the Carlsberg Ridge system, Indian Ocean. *Marine Geology*

Cramp A, Rasul N & **Kenyon N** (1994 in press) Late quaternary deep-sea submarine fane facies: Examples from the Indus system. *Sedimentology*

Draper S (1988) The geotechnical properties of biogenic carbonate ooze from the Rodrigues Ridge, western Indian Ocean. MSc Thesis, University College of North Wales, Bangor.

Elliott AJ & **Savidge G** (1990) Some features of the upwelling off Oman. *Journal of Marine Research* **48** pp 319-333

Evans RL (1991) Electrical resistivity structure of the East Pacific Rise near 13°N. PhD Thesis, University of Cambridge.

Evans RL, Constable SC, Sinha MC, Cox CS & **Unsworth MJ** (1991) Upper crustal resistivity structure of the East Pacific Rise near 13°N. *Geophys. Res. Lett.* **18**. pp 1917 - 1920.

Evans RL, Sinha MC, Constable SC & **Unsworth MJ** (1992) On the electrical nature of the axial melt zone at 13°N on the East Pacific Rise. *J. Geophys. Res.* (submitted Sept 1992)

Fine RA (1993) Circulation of Antarctic Intermediate Water in the South Indian Ocean. *Deep Sea Research I.* Vol **40**. No 10, pp2021-2042

Hardy NC (1991) A marine geophysical study of the Pacific margins of Colombia and southeast Panama. PhD Thesis. University of Birmingham. 253p

Hardy NC (1991) Tectonic evolution of the easternmost Panama Basin: some new data and inferences. *Journal of South American Earth Sciences*. Vol **4**, No 3. pp 261-269

Hardy NC, Westbrook FK & **Heath RP** (1994 in press) The convergent Pacific continental margin of Colombia and eastern Panama. Submitted to *Marine and Petroleum Geology*.

Heath RP (1994) The development of a system for the processing and interpreation of side-scan sonar data, and its application to GLORIA data from the Pacific margin of Colombia and Panama. Unpublished PhD Thesis. 188pp

Hermelin JOR & **Shimmield GB** (1990) The importance of the oxygen minimum zone and sediment geochemistry in the distribution of recent benthic foraminifera in the northwest Indian Ocean. *Marine Geology* **91**. pp 1-29

Hermelin JOR & **Shimmield GB** (1994 in press) The impact of productivity events on benthic foraminiferal fauna in the Arabian Sea. *Palaeoceanography*.

Heward G (1987) Recent foraminifera from the waters around the island of Rodrigues, southern Indian Ocean. MSc Thesis. University of Hull

Heywood KJ, Barton ED & **Simpson JH** (1990) The effect of flow disturbance by an oceanic island. *Journal of Marine Research* **48**, pp 55-73

Heywood KJ, Scrope-Howe S & **Barton ED** (1991) Estimation of zooplankton abundance from shipborne ADCP backscatter *Deep Sea Research* Vol **38** No 6, pp 677- 691

Heywood KJ, Barton ED & **Allen GL** (1994 in press) Observations of 50-day oscillation in currents of the western Indian Ocean *Oceanologica Acta*

Hitchcock GL & **Olson DB** (1992) NE and SW monsoon conditions along the Somali coast during 1987. *Oceanography of the Indian Ocean*. pp 583-593

Hogg NG (1991) Mooring motion corrections revisited. *Journal of Atmospheric and Oceanic Technology*. Vol **8** No 2. pp 289-295

Hogg NG (1992) On the transport of the Gulf Stream between Cape Hatteras and the Grand Banks. *Deep-Sea Research*. Vol **39**, No 7/8. pp 1231-1246

Hogg NG (1994 in press) Observations of Gulf Stream meander-induced disturbances. Accepted by *Journal of Physical Oceanography*. 28pp + figures

Johnson GC, Warren BA & **Olson DB** (1991) Flow of bottom water in the Somali Basin. *Deep Sea Research* Vol **38**, No 6. pp 637-652

Johnson GC, Warren BA & **Olson DB** (1991b) A deep boundary current in the Arabian Basin. *Deep Sea Research* Vol **38**, No 6. pp 653-661

Kearns EJ ***et al***, (1989) CTD and bottle data from RRS *Charles Darwin*, MASAI 1986-1987. University of Miami and WHOI Data Report

Kenyon NH (1987) RRS *Charles Darwin* Cruise 20, 31 January - 27 February 1987. GLORIA Study of the Indus Fan. Institute of Oceanographic Sciences Deacon Laboratory Cruise Report. No **198**. 17pp

Kenyon NH, Ayub A & **Cramp A** (1995) Geometry of the younger sediment bodies of the Indus Fan. Atlas of Deep Water Environments: Architectural style in turbidite systems. Chapman and Hall, London. pp 89-93

Kenyon NH, Ayub A & **Cramp A** (1994 in prep) Morphology and growth patterns of a channel-levee complex and sand lobes on the Indus deep-sea fan.

Khan AA (1989) Geochemistry and palaeoclimate changes in sediments: northern Arabian Sea. PhD Thesis. University of Edinburgh

King BA & **Webb DJ** (1990) Near surface observations of the hydrography and currents of the western equatorial Pacific Ocean. Tropical Oceans Global Atmosphere: abstracts. International TOGA Scientific Conference 16-20 July 1990, Honolulu

King BA ***et al*** (1991) SeaSoar data from the western equatorial Pacific Ocean collected on RRS *Charles Darwin* Cruise 34A September 1988. Institute of Oceanographic Sciences Deacon Laboratory Report No **291** 89pp

Livermore RA, Tomlinson JS & **Woollett RW** (1991) Unusual sea-floor fabric near the Bullard fracture zone imaged by GLORIA sidescan sonar. *Nature*.Vol **353**. pp 158-161

Mantoura RFC (1991) Nutrient and trace metal biogeochemical sampling in the north-western Indian Ocea. RRS *Charles Darwin* Cruise CD 16/86 8 September-12 October 1986, PML Cruise Report

Mantoura RFC, Law CS, Owwens NJP, Burkill PH, Woodward EMS, Howland RJM, & **Llewellyn CA** (1993) Nitrogen biogeochemical cycling in the northwestern Indian Ocean. *Deep Sea Research II*. Vol **40** No 3. pp 651-671

Masson DG (1988) Active margin tectonics in eastern Indonesia; a study with GLORIA and underway geophysics. Institute of Oceanographic Sciences Deacon Laboratory Report No **202**. 20pp

Masson DG, Milsom J, Barber AJ, Sikumbang N & **Dwiyanto B** (1991) Recent tectonics around the island of Timor, eastern Indonesia *Marine and Petroleum Geology*. Vol **8**. pp 35-50

Masson DG, Parson LM, Milsom J, Nichols G, Sikumbang N, Dwiyanto B & **Kallagher H** (1990) Subduction of seamounts at the Java Trench: a view with long-range sidescan sonar. *Tectonophysics* **185**. pp 51-65

McNeill GW & **Shimmield GB** (1991) Diagenetic controls on uranium, molybdenum and vanadium enrichment in organic-rich marine shelf sediments. Heavy metals in the Environment - report of an international conference. pp 436-439

McNeill G W (1993 in prep) The geochemical palaeoceanography and mineralogy of marine sediments for the Peruvian continental margin.

Miles PR & **Roest WR** (1993) Earliest sea-floor spreading magnetic anomalies in the north Arabian Sea and the ocean-continent transition. *Geophys. J. Int.* **115**, pp 1025-1031

Milsom J, Masson DG & **Nicols G** (1992) Three trench endings in eastern Indonesia Marine *Geology* **104** pp 227-241

Minshull TA (1989) Multi-channel seismic studies of sediment accretion and anomalous fracture zone crust. PhD Thesis. University of Cambridge.

Minshull TA & **White RS** (1989) Sediment compaction and fluid migration in the Makran accretionary prism. *Journal of Geophysical Research* **94**, B6. pp 7387-7402

Minshull TA, White RS, Barton PJ & **Collier JS** (1992) Deformation at plate boundaries around the Gulf of Oman. *Marine Geology* **104**. pp 265-277

Minshull TA, Singh SC & **Westbrook GK** (1994) Seismic velocity structure at a gas hydrate reflector offshore western Colombia, from full waveform inversion. *Journal of Geophysical Research.* Vol **99,** No B3. pp 4715-4734

Mitchell NC & **Parson LM** (1993) The tectonic evolution of the Indian Ocean Triple Junction, Anomaly 6 to present. *Journal of Geophysical Research*. Vol **98**, No B2. pp 1793-1812

Molinari RL, Olson DB & **Reverdin G** (1990) Surface Current distributions in the tropical Indian Ocean derived from compilations of surface buoy trajectories. *Journal of Geophysical Research* Vol **95**, No C5 pp 7217-7238

Morales RA & **Barton ED** (1994 In Prep) Variability of water masses in the west Indian Ocean. To be submitted to *Deep Sea Research.*

Morley NH, Statham PJ & **Burton JD** (1993) Dissolved trace metals in the southwestern Indian Ocean. *Deep Sea Research I* Vol **40** No 5. pp1043-1062

Morrison JM & **Olson DB** (1993) Seasonal basinwide extremes in T-S characteristics in the near surface waters of the Arabian Sea and Somali Basin. Oceanography of the Indian Ocean. pp 605-616

Nichols G, Hall R, Milsom J, Masson DG, Parson LM, Sikumbang N, Dwiyanto B & **Kallagher H** (1990) The southern termination of the Philippine Trench. *Tectonophysics* **183**. pp289-303

Olson DB, Hitchcock GL, Fine RA & **Warren BA** (1993) Maintenance of the low-oxygen layer in the central Arabian Sea. *Deep Sea Research II.* Vol **40** No 3 pp 637-685.

Parson LM (1987) RRS *Charles Darwin* Cruise CD 23/87 13 May - 11 June 1987 Geophysical investigation of thee Indian Ocean Triple Junction. Institute of Oceanographic Sciences Deacon Laboratory Report No **201**

Parson LM (1989a) RRS *Charles Darwin* Cruise CD 33/88 5 May - 1 June 1988 Geophysical and geological investigations of the Lau back-arc basin, SW Pacific. Institute of Oceanographic Sciences Deacon Laboratory Report No **206**

Parson LM (1989b) GLORIA long range sidescan sonar results from the Lau Basin, SW Pacific: their use in understanding the ridge geometry and tectonic evolution of active back-arc basins. (in) Geological - Geophysical mapping of the Pacific Region. International Symposium. Yuzhno-Sakhalinsk September 11-19 1989

Parson LM, Hawkins JW & Hunter PM (1992) Morphotectonics of the Lau Basin seafloor - implications for the opening history of backarc basins. Proc. of the Ocean Drilling Program, Initial reports. Vol **135**. pp 81-82

Parson LM & Hawkins JW (1994) Two-stage propagation and the geological history of the Lau back-arc basin. Proceedings of the Ocean Drilling Program, Scientific Results. Vol **135**. 16pp

Parson LM, Patriat P, Searle RC & Briais AR (1993) Segmentation of the Central Indian Ridge between 12°S 12'S and the Indian Ocean Triple Junction. *Marine Geophysical Researches*. **15** pp 265-282

Parson LM, Pearce JA, Murton BJ, Hodkinson RA & RRS ***Charles Darwin*** **Scientific Party.** (1990) Role of ridge jumps and ridge propagation in the tectonic evolution of the Lau back-arc basin, southwest Pacific. *Geology,*. v **18** pp 470-473

Patience AJ (1994 in prep) Geochemical indicators of palaeo-productivity and palaeoclimatology in eastern equatorial Pacific Islands

Pattiaratchi CB & Collins MB (1988) Currents in submarine canyons and their implications for sediment transport. Australian Marine Sciences Association. Jubilee Conference 1963-88. 6pp

Pattiaratchi CB, Poulos S & Collins MB (1994 in prep) Currents in sub-marine canyons; Eastern Mediterranean Sea.

Pedersen TF, Shimmield GB & Price NB (1992) Lack of enhanced preservation of organic matter in sediments under the oxygen minimum on the Oman Margin. *Geochimica et Cosmochimica Acta*. Vol **56**. pp 545-551

Poulos S, Pattiaratchi CB, Cramp A & Collins MB. (1994 in prep) Oceanographic and sedimentological observations along the Hellenic Trench, Ionian Sea.

Prunier KT (1992) The spreading, mixing and ventilation of thermocline and intermediate waters in the Arabian Sea. *Deep Sea Research* (in press). 42pp

Rasul N (1992) Late Quaternary to Present Day coarse grained sediment of the Indus fluvial-marine system. PhD Thesis. University of Wales

Redbourn LJ, Bull JM, Scrutton RA & Stow DAV (1993) Channels, echo character mapping and tectonics from 3.5 kHz profiles, distal Bengal Fan. *Marine Geology* **114**. pp 155-170

Richards KJ & Pollard RT (1991) Structure of the upper ocean in the western Equatorial Pacific. *Nature* **350**, pp 48-50.

Richards KJ *et al*, (1988) RRS *Charles Darwin* Cruise 32, 3 April - 2 May 1988. A study of the upper density and current structure off the western equatorial Pacific. Southampton University, Department of Oceanography. 19pp

Rusby RI (1992) Tectonic Pattern and Evolution of the Easter Microplate, based on GLORIA and other geophysical data. PhD Thesis. University of Durham

Rusby RI & Searle RC (1993) Intraplate thrusting near the Easter microplate. *Geology*. Vol **21**. pp 311-314

Rusby RI & Searle RC (1995 in press) A history of the Easter Microplate 5 Ma to Present. *J.Geophysics Res.*

Savidge G, Elliott AJ & Hubbard LML. (1988) Report of the Oceanographic Survey of the Southern Oman Coast August 1987. (Submitted to the Council for the Conservation of the Environment and Water Resources, Sultanate of Oman and Regional Organisation for the Protection of the Marine Environment, Kuwait) Queen's University of Belfast Marine Biology Station. 52pp

Scrutton RA (1988) RRS *Charles Darwin* Cruise 28 - Cruise Report University of Edinburgh 16pp

Searle RC, Rusby RI, Engeln J, Hey RN, Zukin J, Hunter PM, LeBas TP, Hoffmann H-J & **Livermore R.** (1989) Comprehensive sonar imaging of the Easter Microplate. *Nature* Vol **341**, No 6244. pp 701-705

Searle RC, Bird RT, Rusby RI & **Naar DF** (1993) The development of two oceanic microplates: Easter and Juan Fernandez microplates, East Pacific Rise. *J. Geol. Soc.* Vol **150**. pp 965-976

Searle RC, Francheteau J & **Cavaglio B** (1995 in prep) New observations on mid-plate volcanism and tectonic history of the Pacific plate, Tahiti to Easter Microplate. Earth and Planetary Science Letters

Shimmield GB, Price NB & **Pedersen TF** (1990) The influence of hydrography, bathymetry and productivity on sediment type and composition of the Oman Margin and in the Northwest Arabian Sea. Geological Society Special Publication No **49**. pp759-769

Shimmield GB (1992) Can sediment geochemistry record changes in coastal upelling palaeoproductivity? Evidence from northwest Africa and the Arabian Sea. Geological Society Special Publication No **64** pp29-46

Sinha MC (1989) Cruise Report - RRS *Charles Darwin* Cruise 34. University of Cambridge. 33pp

Sinha MC & **Constable S** (1989) RRS *Charles Darwin* Cruise 39 - Preliminary Report. University of Cambridge. 4pp

Sinha MC, Patel PD, Unsworth MJ, Owen TRE & **MacCormack MRG** (1990) An active source electro-magnetic sounding system for marine use *Mar. Geophys. Res.* **12**. pp 59-68
Tarbell SA, Worrilow SJ & **Hogg NG** (1992) A Compilation of

Moored Current Meter Data from SYNOP Arrays One and Two (September 1987 to July 1989) Volume **XLIV**. Woods Hole Oceanographic Institution. WHOI -93-01. 75pp

Tomlinson JS, Pudsey JC, Livermore RA, Larter RD & **Barker PF**. (1992) Long-range sidescan sonar (GLORIA) survey of the Antarctic Peninsula Pacific margin. Recent Progress in Antarctic Earth Science. Terra Scientific Publishing Company. pp 423-429

Toole JM & **Warren BA** (1993) A hydrographic section across the subtropical South Indian Ocean. *Deep Sea Research I* Vol **40**, No 10. pp 1973-2019

Unsworth MJ (1991) Electro-magnetic exploration of the oceanic crust with controlled sources. PhD Thesis, University of Cambridge.

von Rad U, Reich V, Wissmann G, Stevenson AJ, Morton JL & **Sinha MC** (1990) Seafloor reflectivity, sediment distribution and magmatic anomalies of the Lau Basin (SW Pacific, SO35/48 Cruises). *Marine Mining* **9**. pp 157-181

Warrren BA (1993) Circulation of north Indian deep water in the Arabian Sea. *Oceanography of the Indian Ocean.* pp 575-582

Warren BA & **Johnson GC** (1992) Deep Currents in the Arabian Sea in 1987. *Marine Geology* **104**, pp 279-288

Watson TS (1989) Sediment geochemistry of the Oxygen minimum zone: northwest Indian Ocean. PhD Thesis. University of Edinburgh.

Webb DJ *et al* (1990) RRS *Charles Darwin* Cruise 34A, 15 August - 30 September 1988. The near surface physical oceanography and meteorology of western equatorial Pacific Ocean. Institute of Oceanographic Sciences Deacon Laboratory Cruise Report No **207** 34pp

Westbrook GK (1989) RRS *Charles Darwin* 36/88 Cruise Report 1-

28 December 1988. University of Birmingham. 16pp

Westbrook GK (1989) RRS *Charles Darwin* 40/89 Cruise Report 19 June - 22 July 1989 University of Birmingham 16pp + Map insert

Westbrook GK & **Hardy NC** (1994 in press) Mechanisms of accretion and tectonic erosion in the accretionary complex of the Pacific margin of Colombia. Submitted to *Tectonics*.

Westbrook GK & **Lothian A** (1994 in prep) Investigation of the tectonics and sedimentation of the Chile Ridge - Chile Trench junction with GLORIA long-range sidescan sonar. 21pp + 14 figures

Westbrook GK, Hardy NC & **Heath RP** (1994 in press) Structure and tectonics of the Panama -Nazca plate boundary, in Geologic and Tectonic Development of the Caribbean plate boundary in southern Central America. Special paper of the Geological Society of America.

White RS (1987) Cruise Report. RRS *Charles Darwin* 18/86. Geophysical investigations in the Gulf of Oman. University of Cambridge. 41pp

Williams CA (1986) Cruise Report RRS *Charles Darwin* 14A/86. University of Cambridge. 6pp

Woellner RA (1989) Modern to late quaternary coccolithophorids as indicators of monsoonal upwelling in the Arabian Sea. MS Thesis. University of Miami.

Worrilow SJ (1989) Cruise Report for RRS *Charles Darwin* Cruise 41.Woods Hole Oceanographic Institution. 9pp

Appendix 1

Cruise Summaries - New Horizons in the Ocean

CRUISE SUMMARIES

Dates	Principal Scientist (Affiliation)	Discipline	Area
	1986		
17 June - 17 July	Prof M Brooks (U of Wales College of Cardiff)	Geophysics	Ionian Sea
12 Aug - 3 Sept	Dr H Elderfield (U of Cambridge)	Geochemistry	W Indian Ocean
8 Sept - 11 Oct	Dr R Mantoura (Plymouth Marine Lab)	Biochemistry	NW Indian Ocean
15 Oct- 7 Nov	Dr N Price (U of Edinburgh)	Geochemistry	NW Indian Ocean
14 Nov - 13 Dec	DR R White (U of Cambridge)	Geophysics	N Indian Ocean
18 Dec - 18 Jan	Prof D Olson (U of Miami)	Physics	Arabian Sea
	1987		
31 Jan - 27 Feb	Dr N Kenyon (IOSDL) (Institute of Oceanographic Sciences Deacon Laboratory)	Geophysics	Arabian Sea
1 March - 7 April	Dr A Baxter (City of London Poly)	Geophysics	Indian Ocean
11 April - 8 May	Dr E Barton (U College of N Wales)	Physics	Indian Ocean
13 May - 11 Jun	Dr L Parson (IOSDL)	Geophysics	Indian Ocean
16 June - 14 July	Dr E Barton (U College of N Wales)	Physics	Indian Ocean
17 July - 14 Aug	Prof D Olson (U of Miami)	Physics	Arabian Sea
17 Aug - 27 Aug	Dr G Savidge (U of Belfast)	Physics	Gulf of Oman
31 Aug - 24 Sept	Dr A Cramp (U College, Swansea)	Geophysics	Indus Cone
27 Sept - 1 Nov	Dr R Scrutton (U of Edinburgh)	Geophysics	E Indian Ocean
12 Nov - 18 Dec	Dr B Warren (Wood's Hole Oceanogrphic Inst)	Physics	Indian Ocean Transect
	1988		
6 Feb - 6 March	Dr D Masson (IOSDL)	Geophysics	Off E. Indonesia
3 April - 2 May	Dr K Richards (U of Southampton)	Physics	S W Pacific Ocean
5 May - 1 June	Dr L Parson (IOSDL)	Geophysics	S Pacific Ocean
11 July - 11 Aug	Dr M Sinha (U of Cambridge)	Geophysics	S Pacific Ocean
24 Aug - 30 Sept	DR D Webb (IOSDL)	Physics	S W Pacific Ocean
13 Oct - 17 Nov	Dr R Searle (IOSDL)	Geophysics	Easter Island
1 Dec - 1 Jan	Prof G Westbrook (U of Birmingham)	Geophysics	Chilean Trench
	1989		
3 Jan - 10 April	Dr P Barker (British Antarctic Survey)	Geophysics	Weddel Sea S Atlantic Ocean
14 April - 11 May	Dr N Price (U of Edinburgh)	Geochemistry	Off Peru
25 May - 16 June	Dr M Sinha (U of Cambridge)	Geophysics	E Pacific Rise
20 June - 22 July	Prof G Westbrook (U of Birmingham)	Geophysics	Off Colombia
24 July - 7 Aug	Dr N Breen (Lamont Dogherty Geol Observ)	Geophysics	Off Colombia
10 Aug - 31 Aug	Dr N Hogg (Wood's Hole Oceanographic Inst)	Physics	Mid-Atlantic Ocean

NEW HORIZONS IN THE OCEAN

Chairman for Morning Session
Dr Charles Fay, Superintendent, RVS

1100	Welcome by Mr Brian Hinde Director Scientific Services		
1115	Keynote Address "The Significance of Global Marine Science" Sir Anthony Laughton FRS		
1145	Planning a Pacific Campaign - the theory Dr Martin Angel, IOSDL		
1210	Planning a Pacific Campaign - the practice Dr Stuart White, NERC HQ		
1230	Around the world in 1200 days - from home Mr Christopher Adams, Operations Officer, RVS		
1250	Lunch		

Chairman for Afternoon Session
Professor Mike Brooks, University College of Wales of Cardiff

1400	Around the world in 1200 days - from abroad Mr George Batten, Chief Engineer, RRS Charles Darwin		
1420	Physical Oceanography	1	Prof Don Olson University of Miami
		2	Dr Des Barton University College of North Wale
1500	Geophysics	1	Prof Graham Westbrook University of Birmingham
		2	Dr Martin Sinha Universityof Cambridge
1540	Tea		
1610	Biology amd Biochemistry	1	Dr Graham Savidge University of Belfast
		2	Dr Brian Price University of Edinburgh
1650	Discussion		
1730	Close		

Appendix 2

Deck Plans - Principal Particulars

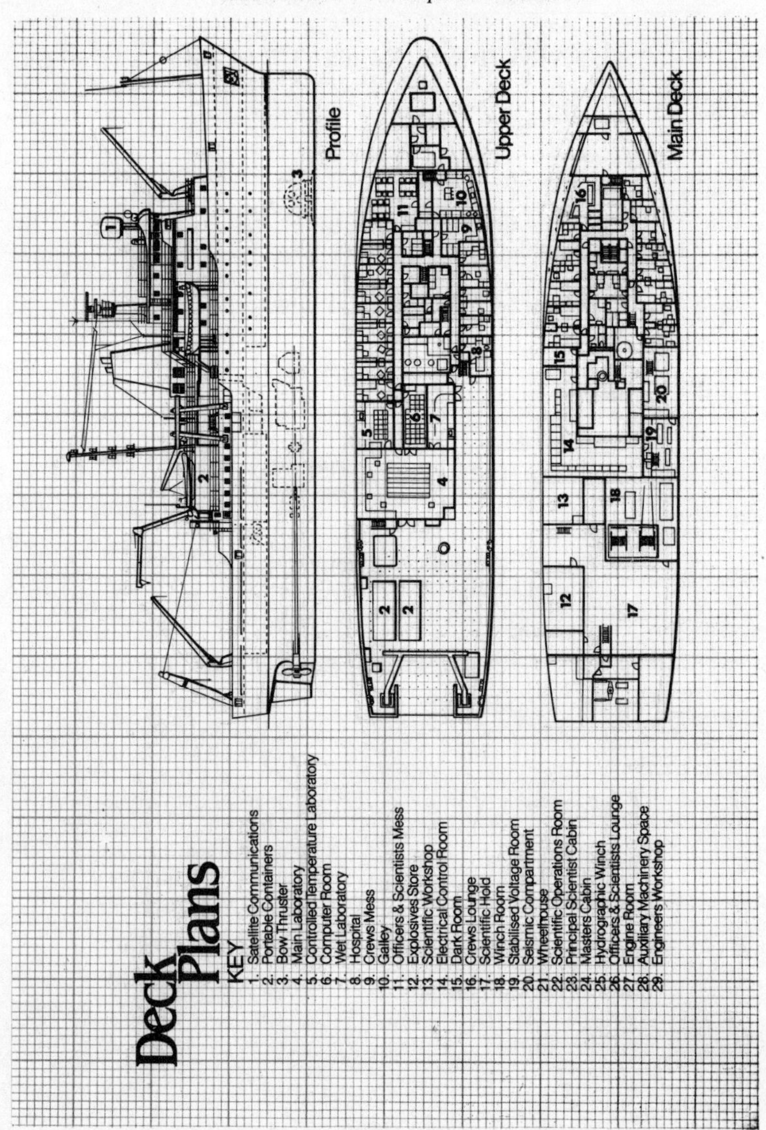

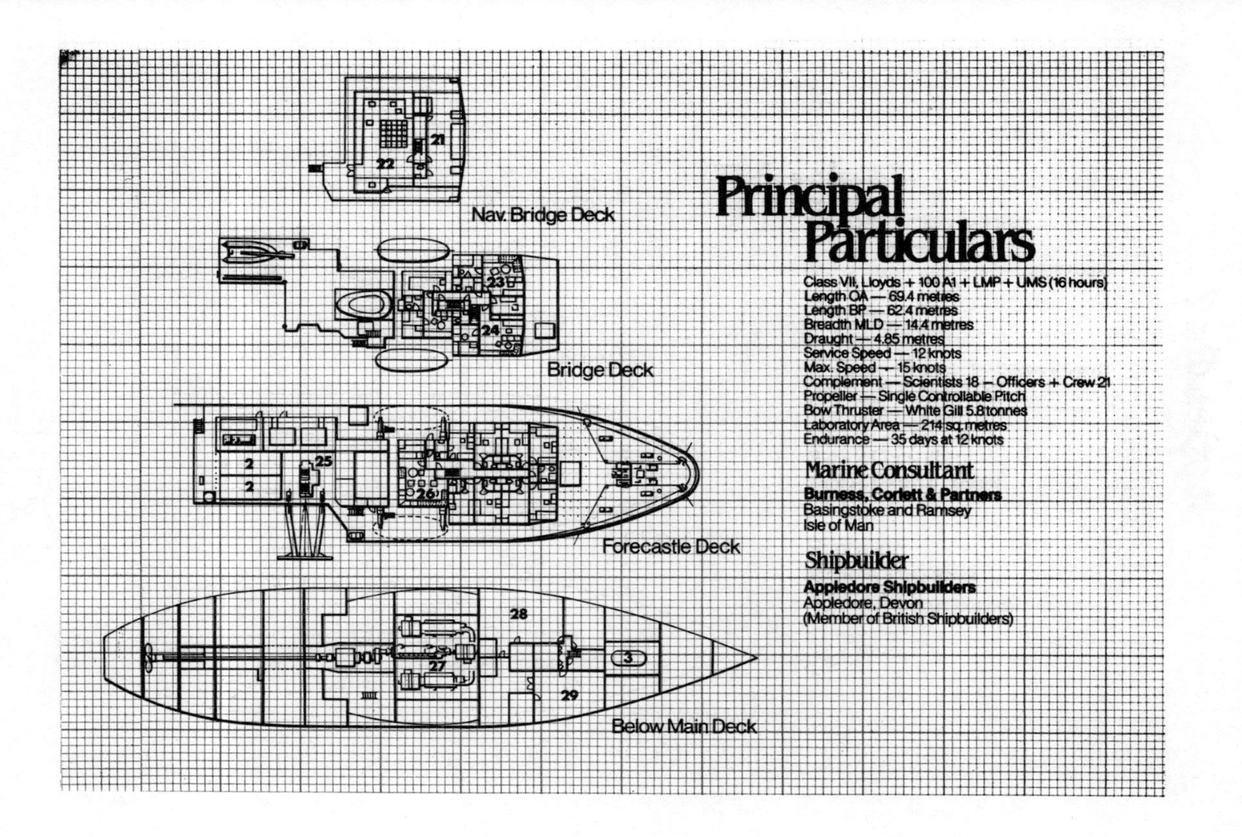
21
22
Nav. Bridge Deck
23
24
Bridge Deck
2
2
25
26
Forecastle Deck
28
27
29
3
Below Main Deck
Principal Particulars
Class VII, Lloyds + 100 A1 + LMP + UMS (16 hours)
Length OA — 69.4 metres
Length BP — 62.4 metres
Breadth MLD — 14.4 metres
Draught — 4.85 metres
Service Speed — 12 knots
Max. Speed — 15 knots
Complement — Scientists 18 — Officers + Crew 21
Propeller — Single Controllable Pitch
Bow Thruster — White Gill 5.8 tonnes
Laboratory Area — 214 sq. metres
Endurance — 35 days at 12 knots
Marine Consultant
Burness, Corlett & Partners
Basingstoke and Ramsey
Isle of Man
Shipbuilder
Appledore Shipbuilders
Appledore, Devon
(Member of British Shipbuilders)

Appendix 3

Facts and Figures

RRS *Charles Darwin* has been away from UK for 1187 days (1200 rounded).
She made 50 port calls in 12 countries (note - not calls at 50 ports - some ports were visited several times).

She has steamed 133,600 nautical miles, and used 4,500 tonnes of fuel (1.2M gallons). Fuel consumption is thus:

either	0.11 miles/gallon (9 gallons/mile)
or	29.3 miles/tonne
or	3.8 (say 4) tonnes/day

The cost of the 3.5 year voyage is about £12M.

The food costs alone are about £165,000.

Finally, if all of the ship's power were used on shore, it could supply a town of about 350 houses - or say the homes of about 1000 people.

Index